Fluid Effects in Polymers and Polymeric Composites

Mechanical Engineering Series

Frederick F. Ling
Editor-in-Chief

The Mechanical Engineering Series features graduate texts and research monographs to address the need for information in contemporary mechanical engineering, including areas of concentration of applied mechanics, biomechanics, computational mechanics, dynamical systems and control, energetics, mechanics of materials, processing, production systems, thermal science, and tribology.

For further volumes:
http://www.springer.com/series/1161

Y. Jack Weitsman

Fluid Effects in Polymers and Polymeric Composites

All corrections and inquiries may be sent to
Dr. Dayakar Penumadu at dpenumad@utk.edu

ISSN 0941-5122 e-ISSN 2192-063X
ISBN 978-1-4899-7331-3 ISBN 978-1-4614-1059-1 (eBook)
DOI 10.1007/978-1-4614-1059-1
Springer New York Dordrecht Heidelberg London

Printed on acid-free paper

Springer is part of Springer Science+Business Media (www.springer.com)

To my wife Berthe

In Memoriam

Yechiel Jack Weitsman

1932–2010

Dr. Yechiel Jack Weitsman was born on January 15, 1932 in Poland and died in Knoxville, Tennessee on June 16, 2010 at the age of 78. Jack dearly loved his wife Bertha, their children Jonathan and Debbie, and his son-in-law Chris. His two grandchildren, Elon and Gideon, were love of his life. Jack is remembered for his sense of humor and dignified dealings with all his students and colleagues.

Jack was a man of intense faith and had unending commitment to his heritage.

Preface

This book on the "Effects of Fluids on Polymers and Polymeric Composites" written by my good friend and colleague, late Dr. Y. Jack Weitsman, is a relatively rare and comprehensive treatise on understanding the degradation of polymer composites and polymeric sandwich structures to harsh environment comprising of high level of humidity or for the case of marine structures which are completely surrounded by sea water. As was demonstrated, most polymeric composites absorb moisture, which is characterized by the weight gain data with time. From such simple measurements, profound statements can be inferred with respect to long-term behavior and stability of such composites when exposed to fluids (mostly water). This book has ten chapters arranged in a systematic fashion.

Chapter 1 provides a summary outlook on this complex topic with suggested three key references that address the static and dynamic mechanical behavior and associated degradation of composites from fluid exposure. Since the topic requires mastery of polymer science and engineering, engineering mechanics, chemical and polymer characterization, and thermodynamics of multi-phase materials, the author's perspective is more slanted on solid mechanics aspect of these materials. Jack had consistent financial support for three decades to study this topic from the Solid Mechanics Program managed by Dr. Yapa Rajapakse at the United States Office of Naval Research (ONR). Thus, the author is uniquely qualified to address this complex topic.

Chapter 2 provides a quick summary of materials involved with composite materials associated polymeric resin, reinforced fibers, and composite systems. Much of the discussion is focused on glass, graphite, and carbon fiber reinforcements and a few select resin systems that have been used in aerospace and naval applications. Thus, the material is of practical and immediate relevance to a broad class of problems.

Chapters 3–5 address the mechanisms associated with fluid ingress and polymer molecular structure, simple ways in the laboratory to obtain phenomenological data for composites prepared using a given manufacturing process (for example using pre-preg's cured in an autoclave oven or recently popular VARTM based approach) using moisture weight gain studies, and interpreting the data with mathematically rigorous diffusion models, reasons for fluid ingress, and coupling of fluid and time effects on mechanical properties through fluid enhanced viscoelasticity approach.

Chapter 7 includes a very important underlying concept of reduced strength and fatigue life for composites when subjected to repeated loading if the fluid is allowed to enter matrix resin while the composite material is subjected to dynamic loading. If the fluid is incompressible water, then water enters the matrix cracks while the composite sample is being subjected to fatigue loading and can significantly reduce the number of cycles to failure under fluid confined conditions.

Chapter 8 addresses our recent work on the topic of sandwich structures which consist of thin polymeric composites sandwiched with a low density core material such as balsa wood or PVC foam. Such materials are of significant recent interest for building light weight and durable ship structures and fluid (sea water in our studies) can have deleterious effect on interface fracture behavior (quantified using pertinent mode critical energy release rate) and shape distortions resulting form one sided exposure to sea water induced expansion. Shear lag theory has been found to adequately describe the shape distortions observed under controlled conditions for carbon-vinyl ester composite facings and PVC foam core based sandwich.

Chapters 9 and 10 deal with anomaly associated with fluid uptake in polymers and composites based on micro-structure characterization work and major conclusions presented in the book.

In my personal opinion, this book is a must read for engineers and scientists associated with using and developing polymeric composites and novel sandwich structural materials. In addition, it will be an excellent introductory graduate level book for an inter-disciplinary course on the mechanical behavior of polymeric composites, sandwich structures, and integrating basic engineering mechanics to study fluid effects. On a personal note, I want to thank Dr. Akawat Siriruk, my former doctoral student and post-doctoral fellow, who has typed this manuscript for Jack while he was battling cancer and did an outstanding job of helping Jack Weitsman to complete this book. Jack dearly loved his wife Bertha and she was instrumental in providing the peace of mind to Jack to complete this book in most challenging health related circumstances.

I do hope that the reader will enjoy the contents and should find the material useful and significant for a long time to come.

Knoxville, TN, USA

Dr. Dayakar Penumadu
Head, Civil and Environmental
Engineering Department
Fred Peebles Professor and JIAM
Chair of Excellence
University of Tennessee, Knoxville

Acknowledgements

My original intent was to write a brief monograph on a topic that occupied much of my attention for nearly 40 years, reporting mainly on the work performed by me and my collaborators. Not being satisfied with the result I decided to expand its content to include the research of others. This opened Pandora's box and transformed a humble intent into a voracious monster. Despite all my efforts it is safe to say that the monster has not been satiated.

My sincere apologies to many authors for my sin of omission; if their worthy contributions are not mentioned in this book this reflects on my own shortcomings and not on the significance of their work.

This book, with all its flaws, would have not been possible without the encouragement, support, and crucial help of many friends and colleagues.

I am indebted to my friend and mentor, Dr. Fred Ling for planting the idea in my mind, and to Dr. Y. Rajapakse, my program manager at the Office of Naval Research for the past 28 years, for his trust, steady financial support, and unrelenting encouragement for this undertaking. All the research by my students, postdoctoral colleagues and myself reported herein was conducted as part of Dr. Rajapakse's program at ONR.

I am beholden to my friend and colleague, Dr. Dayakar Penumadu of the University of Tennessee for a most agreeable collaboration during the past several years. This has been a very special experience for me and a wonderful way to cap my career.

But my most special gratitude is due to my colleague, dear friend and last Ph.D. student Akawut "O" Siriruk. Without his help in organizing this book, typing its entire text and the setting of both figures and references – there would have been no book.

Dr. Siriruk has also been a source of great personal support during my ongoing illness.

Contents

Chapter 1
Introduction

The subject of this book, "fluid effects on polymers and polymeric composites," draws upon several scientific and technical disciplines. Of prime importance are Polymer Science, Mechanics and Thermodynamics. Of those, Mechanics – which concerns most engineering applications – has been most familiar to the author. Nevertheless, an effort was made to achieve some semblance of balance by incorporating ideas and basic notions of Polymer Science into the text.

The book is in no sense complete and despite all efforts it does not encompass all the relevant work on the subject that abounds in the technical and scientific literature.

A major cause for the aforementioned drawback is the multitude of technologies that relate to fluid effects. Besides the familiar aerospace and naval applications, those effects are of concern for the civil infrastructure, electronics industry, bioengineering and the adhesives industry-to name a few. It follows that articles are scattered over a multitude of journals and a comprehensive review of source material is, most likely, beyond the capability of a single author.

The reader may be helped by referring to several review articles listed at the end of this chapter. Note that the list is focused on mechanistic aspects. A substantial number of reviews centered on polymer science is also available. Regarding mechanistic aspects the reader may note two reviews written by the author (Weitsman 1991, 2000), which contain a large number of references, but by far the most comprehensive review article on the subject (Bond and Smith 2006) contains critical evaluations of referenced articles, in addition to a most considerable reference list. Note also that a list of review articles was included in a previous work (Weitsman 2000).

It is important to note that, despite its vintage, the book on diffusion written by Crank (1980) still contains an irreplaceable treasure trove of information on this subject.

In a broad sense, the majority of the works concerning fluid effects on polymers and polymeric composites focused on data collection, with meager attempts to provide a technical or science based rationale for observed behavior. This book

Y.J. Weitsman, *Fluid Effects in Polymers and Polymeric Composites*,
Mechanical Engineering Series, DOI 10.1007/978-1-4614-1059-1_1,

contains several formulations that attempt to relate several empirically observed phenomena to possible underlying causes.

While striving to include the multitude of results obtained by the large number of researchers in the field, the reader may detect a certain bias toward the work conducted by the author and his many students and post doctoral colleagues.

A brief outline of the book follows:

Chapter 2 contains some selected topics, including several that are less generally familiar, that are to be amplified and utilized in the sequel. These include aspects of composite materials and viscoelasticity.

Chapter 3 discusses various forms of water ingress into polymers and composites, beginning with basic aspects of polymeric molecular structure.

Chapter 4 contains tables of fluid ingress data, attempting to sort those into several categories and interpret their interrelations. Note that although this is the main motivation for several tables, those same lists can be used to gather additional information about the polymers and composites listed therein.

Chapter 5 elaborates on diffusion models, beyond the more elementary scope of Chap. 3. In addition, this chapter provides some science-based formulations for several underlying causes of fluid ingress.

Chapter 6 focuses on the fluid enhanced time-dependent response of polymers and composites together with a comprehensive model for fluid enhanced viscoelasticity.

Chapter 7 concerns the effects of fluids on the strength and fatigue of polymeric composite and provides data and fracture mechanics-based interpretation on the fatigue of fiber reinforced composites.

Chapter 8 focuses on a fairly recent aspect of fluid ingress, namely in conjunction with polymeric foams and polymeric sandwich structures. The recent interest in this issue was reignited in relation to naval applications.

Chapter 9 concerns three specific issues. The first of those is an attempt to explain a long standing anomaly regarding fluid uptake in polymers and composites. This attempt derives from more recent discoveries regarding the microstructure of polymers.

Chapter 10 lists some major conclusion that may be drawn from the information presented in this book, with all due caveats.

References

Bond DA, Smith PA (2006) Modeling the transport of low-molecular-weight penetrants within polymer matrix composites. Appl Mech Rev 59(5):249–268

Crank J (1980) The mathematics of diffusion, 2nd edn. Oxford University Press, Oxford

Weitsman YJ (1991) Moisture in composites: sorption and damage. In: Reifsnider KL (ed) Fatigue of composite materials. Elsevier, New York, pp 385–429

Weitsman YJ (2000) Effects of fluids on polymeric composites–a review. In: Kelly A, Zweben C (editors in Chief) Comprehensive composite materials. Talreja R, Manson J-AE (editors). Polymeric matrix composites, vol. 2. Amsterdam: Elsevier, pp 369–401

Chapter 2
Background to Polymers and Composites

2.1 Fibers

In view of their manufacturing process, such as drawing and extrusion for glass fibers and the sheet-like atomic structure of carbon and graphite fibers, these reinforcements are essentially transversely anisotropic about their longitudinal orientation (Hull 1981), in which direction they possess high stiffness and strength. In view of their small diameter, typically 8 μm, it has been possible to determine their longitudinal modulus only, while no direct data are available about their in-plane stiffnesses.

Typical values of experimental data are listed in Table 2.1

It is common practice to subject fibers to two kinds of treatments. The first one, called "sizing" aims at the polishing and smoothing of nicks and roughnesses over the fibers' surfaces. The second treatment concerns the introduction of an interphase, usually a fraction of a micron thick, whose purpose is to enhance the chemical/physical binding between the polymeric binder and the embedded fibers.

2.2 Polymers

2.2.1 Material Aspects

Polymers are a large class of materials consisting of long chain molecular structures, up to 0 (10^6) in molecular weights, where the basic component of these chains is a monomer. Consider for example the styrene monomer, where two carbon molecules are attached to each other by a double bond. Upon introducing a free radical catalyst, the double bond is opened, releasing a free electron and rearranging the bonding configuration as sketched below (Fig. 2.1).

Y.J. Weitsman, *Fluid Effects in Polymers and Polymeric Composites*,
Mechanical Engineering Series, DOI 10.1007/978-1-4614-1059-1_2,

Table 2.1 Longitudinal properties of some fibers

Fiber	E glass	S glass	Carbon AS4	Carbon T300
Longitudinal stiffness (GPa)	72	85	235	228
Longitudinal strength (MPa)	3,450	4,500	3,750	3,000
Longitudinal elongation to failure (%)	4.8	5.3	1.6	1.3
Density (g/cm^3)	2.55	2.5	1.8	1.75

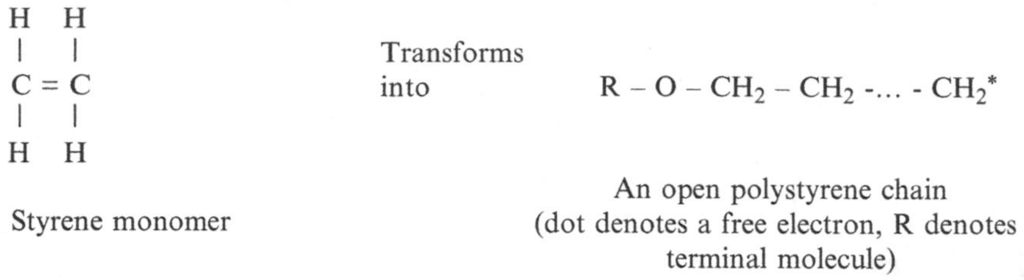

H H
| |
C = C
| |
H H

Styrene monomer

Transforms into

R – O – CH_2 – CH_2 -... - CH_2^*

An open polystyrene chain (dot denotes a free electron, R denotes terminal molecule)

Fig. 2.1 The polymerization of polystyrene

The chain is terminated by linking the last CH_2 component to another free radical. The above scheme represents one form of a polymerization process.

In view of the aforementioned imperfections in the cross-linking process the glassy state still contains an overall amount of free volume $V_f \sim 0.025 - 0.03$, which is distributed throughout the polymer (Deng and Jean 1993; Dammert et al. 1999; Yampolskii 2007). It has been suggested that these micro volumes are located about chain ends.

A typical record of volumetric increase due to absorption of moist air is given in Fig. 2.2, where the gap between the data and the theoretical line accounts for water molecules that occupy the free volume.

It was noted that the glassy phase is not in a state of thermodynamic equilibrium and chain motions, though inhibited, still persist. These motions are not limited to the side groups alone, but involve several mechanisms within the backbone structure as well (Aklonis and MacKnight 1983). The above fluctuations are further arrested at a temperature ranging about $T_\beta < T_g$ where the polymer undergoes a secondary transition (The β transition).

Separately from the effects of temperature, the free volume V_f tends to diminish spontaneously with time toward a more compacted configuration. In that configuration chain mobility is further inhibited and creep is diminished. This tendency, called "aging," is associated with embrittlement and reduction in polymer fracture toughness (Struik 1978).

The activity a_i of the ith component in a vapor mixture is defined by the ratio $a_i = P_i/P_T$, where P_i is its partial vapor pressure and P_T is the total pressure of the vapor mixture. For humid air, a_i of the water vapor is akin to relative humidity (RH) (Bond and Smith 2006). Note that RH is related to vapor pressure in tabulated form. The activity a is a zmeasurable quantity and theoretically predictable (Flory 1953).

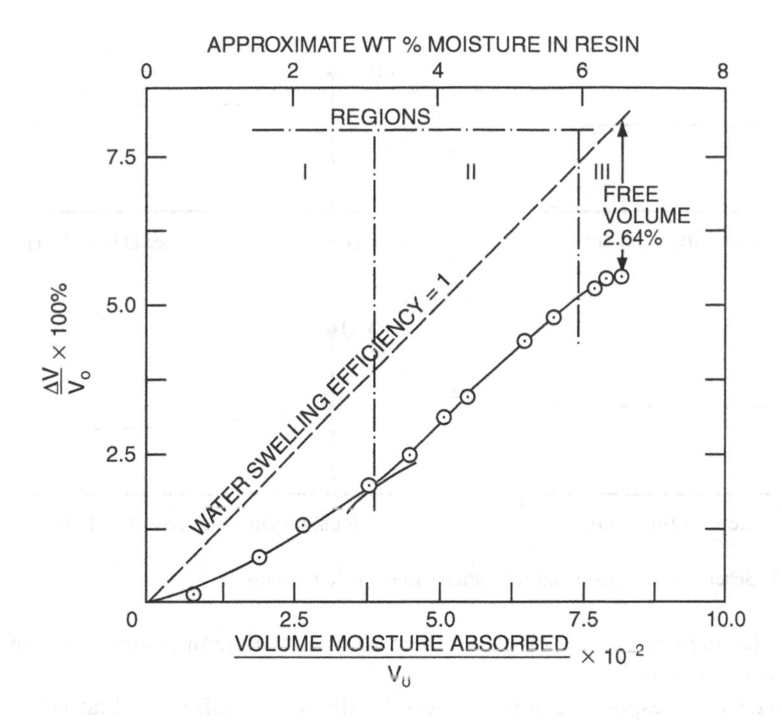

Fig. 2.2 Swelling behavior of Hercules 3501 epoxy resin immersed in water at 74°C. Swelling efficiency = 1 if the volumetric strain of the epoxy matches the volume of absorbed water. With kind permission from Springer Science + Business Media: Adamson (1980) figure 2

2.2.2 *Mechanics Aspects of Glassy Polymers*

In view of the presence of the free volume V_f, the macro-molecular polymeric chains still retain a certain degree of freedom of motion that allows for some configurational changes under loads. These changes may be viewed as occurring against viscous resistance and are therefore time dependent.

Consequently, the strain response to the application of, say, a unit step load is related by a time dependent creep compliance $D(t)$. Inversely, the stress associated with the application of unit strain is given by the relaxation modulus $E(t)$. These are shown schematically in Fig. 2.3.

When excited by thermal agitation, both creep and recovery quicken with temperature. In many cases, thermal effects are focused on time alone and can be accounted for by means of a temperature dependent shift factor $a_T(T)$, thereby $D(t) \rightarrow D(\xi)$, where $\xi = \xi(T) = t/a_T(T)$ and analogously for $E(t)$. This states that the time-dependent response at all temperatures is essentially the same,

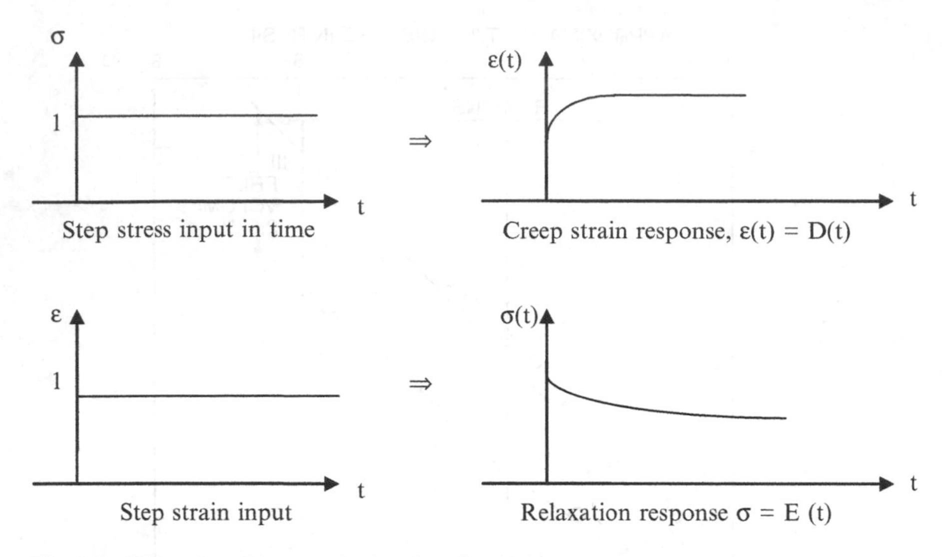

Fig. 2.3 Schematics of creep and relaxation viscoelastic response

except that the gears on the clock are to be adjusted with temperature to cause it to run faster or slower.[1]

Viscoelastic response exhibits reversal effects that differ qualitatively from elastic behavior. Consider, for instance, the following cases (Fig. 2.4).

2.3 Effects of Fluids: Fundamental Aspects

When polymers are exposed to ambient fluids or vapor, the highly mobile and generally small molecules of the ambient phase penetrate the polymer by means of what may be conceived as a random walk process. This process proceeds until the attainment of equilibrium. Equilibrium may be formally attributed to a balance in internal and ambient chemical potentials in the absence of flux.[2] Alternately, it is possible to associate equilibrium with the state where fluid molecules became

[1] It should be noted that $a_T(T)$ is obtained from isothermal creep data, plotted against log t and collected over a wide temperature range. These isothermal plots are then coalesced to form a "master curve" by shifting parallel to the horizontal log t axis. This procedure is not error free. To verify its validity, it is necessary to conduct transient temperature tests and compare the resulting data with computational predictions. For temperatures sufficiently below T_g the shift factor function $a_T(T)$ is given by the Arrhenius expression of the form $a_T(T) = A \exp(-B/T)$.

[2] The chemical potential μ, which is conjugate to the fluid concentration m, is a measure of the energy required the increase the concentration while holding other thermodynamic variables constant. Originally introduced by Gibbs, it was defined by means of the Gibbs free energy function G, i.e., $\mu = \partial G/\partial m$. But other forms of energy can be used.

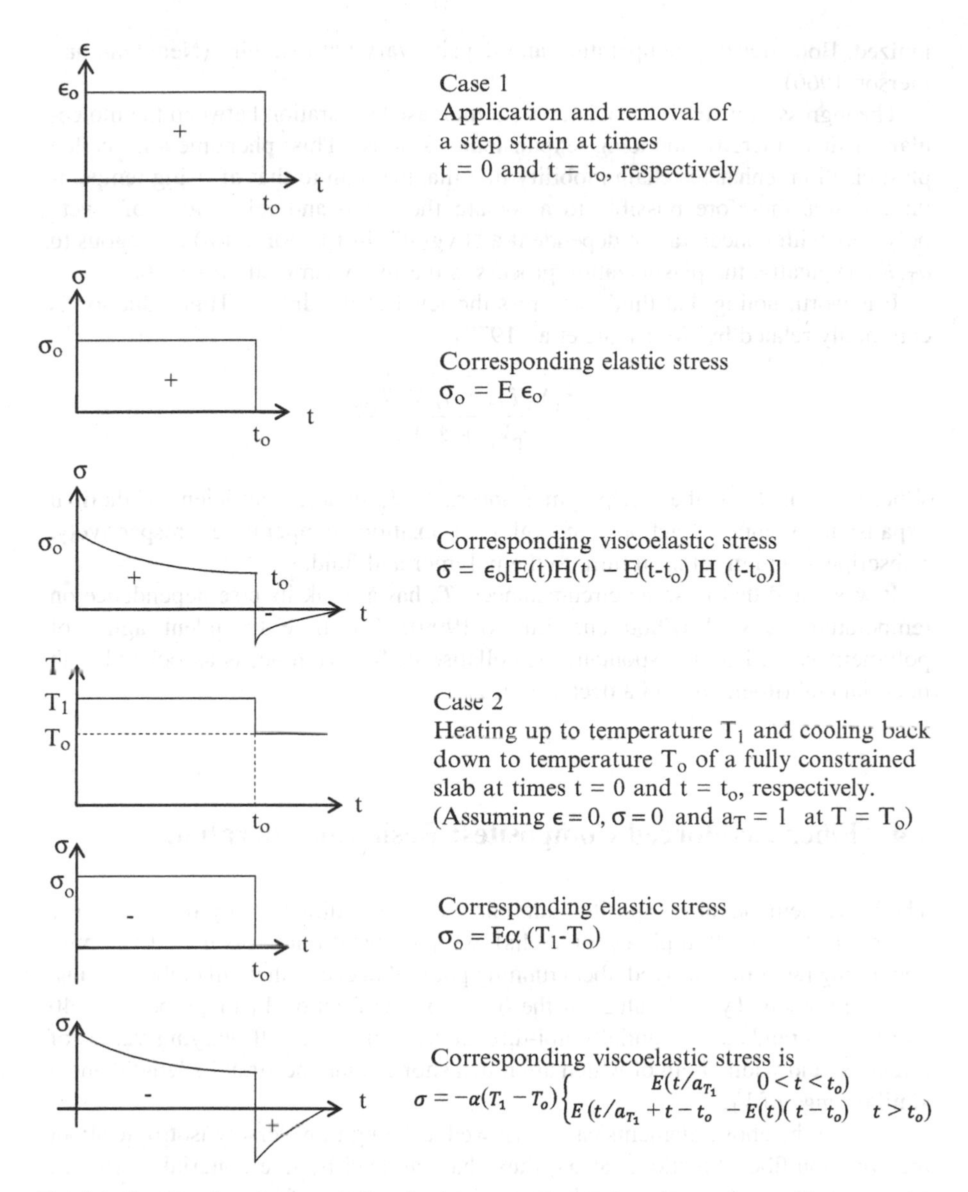

Fig. 2.4 Comparative elastic and viscoelastic responses to strain and stress inputs. Sign reversals occur in the viscoelastic case

attached to all available chemical sites within the polymeric chains and filled to capacity all the accessible free volume within the polymer.

Regarding sea water, this fluid consists of about 3.5% of chemical elements, of which 30% are sodium and 55% chloride. Most of these elements are nearly entirely ionized. Both freezing temperature and density vary with salinity (Neumann and Pierson 1966).

The ingress of water is associated with increased separation between the molecular chains, thereby inducing expansional strains. This phenomenon, called plasticization, enhances chain mobility in a manner akin to that of rising temperature. It was therefore possible to associate the creep and relaxations of "wet" polymers with concentration dependent a "hygral" shift factor $a_H(m)$ analogous to $a_T(T)$. Typically, the plasticization process is thermodynamically reversible.

It is worth noting that fluids suppress the level of the dry T_g. This reduction is commonly related by (McKague et al. 1978)

$$T_{gw} = \frac{\alpha_p V_p T_{gp} + \alpha_m V_m T_{gm}}{\alpha_p V_p + \alpha_m V_m},$$

Where, T_{gw} is T_g of the wet polymer and α, V, T_g denote coefficient of thermal expansion, volume fraction, and glass transition temperature, respectively. Subscripts p and m refer, in turn, to the polymer and fluid.

It was noted that in some circumstances, T_g has a weak inverse dependence on temperature as well (Zhou and Lucas 1999). The time-dependent aging of polymers, caused by the spontaneous collapse of free volume, is associated with material embrittlement and a decrease in T_g.

2.4 Fiber Reinforced Composites: Basic Considerations

The basic elements of composite materials are the uni-directionally reinforced ply or strand. Plies are thin plates (or shells), with a typical thickness $h = 0.125$ mm, containing randomly arrayed fibers running parallel to each other, all embedded in a polymeric resin. Typical values of the fiber volume fraction $\bar{V}_f$ range between 40 and 60%. Strands are essentially uni-directional elements, with varying values of thickness and width, both of which are much shorter than the strand's length, and a similar range of $\bar{V}_f$.

Each of the above elements can be viewed as being transversely isotropic about the common fiber direction, say x_1, thus characterized by five material constants, say E_{11}, E_{22}, G_{23}, G_{12}, and ν_{12}, where $x_2 - x_3$ is the plane of isotropy and ν_{12} is the Poisson's ratio that expresses the strain $-\varepsilon_2/\varepsilon_1$, due to σ_1. For in-plane problems, say within the $x_1 - x_2$ plane, G_{23} is globally irrelevant. Similarly, only two thermal and two hydral expansional coefficients are of interest, namely α_1, α_2 $(=\alpha_3)$ and β_1, β_2 $(=\beta_3)$, respectively.

Typical values of material properties are given in Table 2.2 for several fiber reinforced polymeric composites

Table 2.1 suggests that the effect of % volumetric water gain is approximately equivalent to that of temperature excursion exceeding 100°C.

Table 2.2 Material properties of several uni-directionally reinforced composites

Property	E-glass/ epoxy	S-glass/ epoxy	Carbon/epoxy A54/3501-6	Carbon/epoxy IM6/SC 1081
V_f (%)	55	50	63	65
Density ρ (g/cm^3)	2.1	2	1.58	1.6
E_{11} (GPa)	3.1	43	142	177
E_{22} (GPa)	8.6	8.9	10.3	10.8
G_{12} (GPa)	3.8	4.5	7.2	7.6
ν_{12}	0.28	0.27	0.27	0.27
α_1 (10^{-6}/°C)	7	5	−0.9	−0.3
α_2 (10^{-6}/°C)	21	26	27	30
β_1 (%)	0	0	0.01	0
β_2 (%)	0.3	0.3	0.3	0.3

β_1 and β_2 are per 1% of volume of water uptake at saturation, i.e., composite added weight gain/composite dry weight

Plies and strands are exceedingly weak in their transverse, in-plane directions. Strength enhancement can be achieved by strand weaving or through across-the-thickness stitching. In addition, considerations of cost and versatility in manufacturing suggest that swirl mat and randomly oriented chopped strand mats may have potential advantages.

Uni-directionally reinforced strands are the essential components in the manufacture of woven composites. In similar manner, the uni-directional reinforced plies are the basic building blocks of laminates. The latter contain a lay-up of several, multi-oriented plies and are formed by the application of a prescribed regime of pressure and temperature within an autoclave or a vacuum press.

A state of plane stress is considered to hold for the in-plane response of thin plates. Consequently, the stress-strain relations for a uni-directionally reinforced ply, with fibers parallel to the x_1 direction, employing truncated notation, read

$$\begin{aligned}\sigma_1 &= Q_{11}\varepsilon_1 + Q_{12}\varepsilon_2,\\ \sigma_2 &= Q_{12}\varepsilon_1 + Q_{22}\varepsilon_2,\\ \sigma_6 &= Q_{66}\varepsilon_6.\end{aligned} \tag{2.1}$$

Where, $Q_{11} = \bar{m}E_{11}, Q_{22} = \bar{m}E_{22}, Q_{12} = \bar{m}\nu_{12}E_{22} = \bar{m}\nu_{21}E_{11}, Q_{66} = G_{12}$, where, $\bar{m} = 1/(1 - \nu_{12}\nu_{21})$ with $\nu_{21} = \nu_{12}(E_{22}/E_{11})$ typically, $1.004 < \bar{m} < 1.015$.

Laminate theory, which derives from expressions for the off-axis stiffnesses when plies are oriented at various angles about common laminate axes, say x and y, is well developed and available in many textbooks.

When subjected to ambient temperature and/or moisture, the individual multi-oriented plies within a laminate cannot expand freely and impose mutual constraints on each other by means of internal residual stresses.

Consider, for example, a symmetric, cross-ply laminate [90°/0°/0°/90°] or, in laminate jargon $[90°/0°]_s$. Let all plies be saturated with moisture (or fluid) at a uniform concentration m.

For the 0° plies, expressions (2.1) read

$$\begin{aligned}\sigma_1^0 &= Q_{11}(\varepsilon_1 - \beta_1 m) + Q_{12}(\varepsilon_2 - \beta_2 m),\\ \sigma_2^0 &= Q_{12}(\varepsilon_1 - \beta_1 m) + Q_{22}(\varepsilon_2 - \beta_2 m).\end{aligned} \tag{2.2}$$

While for the 90° plies one has

$$\begin{aligned}\sigma_1^{90} &= Q_{12}(\varepsilon_2 - \beta_1 m) + Q_{22}(\varepsilon_1 - \beta_2 m),\\ \sigma_2^{90} &= Q_{11}(\varepsilon_2 - \beta_1 m) + Q_{12}(\varepsilon_1 - \beta_2 m).\end{aligned} \tag{2.3}$$

In the absence of applied stress $\sigma_1^0 + \sigma_1^{90} = 0$ and $\sigma_2^0 + \sigma_2^{90} = 0$. Thereby, it follows that

$$(Q_{11} + Q_{22})\varepsilon_1 + 2Q_{12}\varepsilon_2 = A(Q;\beta)m,$$

$$2Q_{12}\varepsilon_1 + (Q_{11} + Q_{22})\varepsilon_2 = A(Q;\beta)m,$$

Where, $A(Q;\beta) = (Q_{11} + Q_{12})\beta_1 + (Q_{12} + Q_{22})\beta_2$.
Thus

$$\varepsilon_1 = \varepsilon_2 = \beta^+ m, \tag{2.4}$$

Where

$$\beta^+ = A(Q;\beta)/(Q_{11} + Q_{22} + 2Q_{12}). \tag{2.5}$$

For most composites $\beta_1 \approx 0$ and $Q_{11} \gg Q_{22} + 2Q_{12}$, thus approximately

$$\beta^+ \sim \beta_2(Q_{12} + Q_{22})/Q_{11}, \text{ i.e., } 0(\beta_2/10). \tag{2.6}$$

A similar result holds for the case of thermal expansion.

References

Adamson MJ (1980) Thermal expansion and swelling of cured epoxy resin used in graphite/epoxy composite materials. J Mater Sci 15(7):1736–1745

Aklonis JJ, MacKnight WJ (1983) Introduction to polymer viscoelasticity, 2nd edn. Wiley-Interscience/Wiley, New York, pp 40–53

Bond DA, Smith PA (2006) Modeling the transport of low-molecular-weight penetrants within polymer matrix composites. Appl Mech Rev 59(5):249–268

Dammert RM, Maunu SL, Maurer FHJ, Neelov IM, Niemela S, Sundholm F, Wastlund C (1999) Free volume and tacticity in polystyrenes. Macromolecules 32(6):1930–1938

Deng Q, Jean YC (1993) Free-volume distributions of an epoxy polymer probed by positron annihilation: pressure dependence. Macromolecules 26(1):30–34

Flory PJ (1953) Principles of polymer chemistry. Cornell University Press, Ithaca

Hull D (1981) Introduction to composite materials. Cambridge University Press, Cambridge

McKague EL Jr, Reynolds JD, Halkias JE (1978) Swelling and glass transition relations for epoxy matrix material in humid environments. J Appl Polym Sci 22(6):1643–1654

Neumann G, Pierson WJ (1966) Principles of physical oceanography. Prentice-Hall, Englewood Cliffs

Struik LCE (1978) Physical aging in amorphous polymers and other materials. Elsevier, New York

Yampolskii YP (2007) Methods for investigation of the free volume in polymers. Russ Chem Rev 76(1):59–78

Zhou J, Lucas JP (1999) Hygrothermal effects of epoxy resin. Part I: the nature of water in epoxy. Polymer 40(20):5505–5512

Chapter 3
Fluid Ingress Processes, Basic and Preliminaries

3.1 An Overview and Basic Scientific Concepts

Polymers form a vast class of materials, much too diverse for catergorization under a single rule, especially regarding their interactions with fluids. Nevertheless, recently developed devices and probing techniques provide illuminating insights in several specific circumstances. In several cases, these methods enabled the judicious distinction between the portions of liquid water and water molecules that were attached by one or more hydrogen bonds to the polymeric chains (Sammon et al. 1998; Marechal and Chamel 1996; Cutugno et al. 2001; Kusanagi and Yukawa 1994).

The total free volume V_f consists of an agglomeration of micro volumes whose specific distribution varies with the polymeric configurations and depends on external pressure. For polypropelene, the dimensions of the micro diameters range between 0.5 and 4.0 Å (Deng and Jean 1993), while in other cases the micro volumes may attain diameters of up to 8 Å. For epoxy, the average pore diameter was estimated to vary between 2.7 and 5 Å (Soles et al. 2000; Soles and Yee 2000). It follows that some of these micro volumes are too small to admit H_2O water molecules whose kinetic diameter varies between 1.5 and 3 Å.

However, it should be borne in mind that polymers in general and epoxies in particular consist of a large family of cross-linked resins with disparate chemical properties.

It is generally agreed that water migrates within polymeric resins to fill the available micro volumes. This migration may occur by diffusion through a process of random walk in combination with configurational changes of the polymeric chains (Liang et al., private communication) or by similar motion through tubular nano-pores distributed within the polymer (Soles and Yee 2000). In some polymers, the water may consist of clustered and gaseous components, while in others in a gaseous phase alone (Kawagoe et al. 1999). The latter case seems to prevail in hydrophobic polymers (Kusanagi and Yukawa 1994).

Y.J. Weitsman, *Fluid Effects in Polymers and Polymeric Composites*,
Mechanical Engineering Series, DOI 10.1007/978-1-4614-1059-1_3,

Water molecules that diffuse between polymeric chains tend to induce swelling, thus increasing interchain spacings and enhancing their freedom of motion. This phenomenon, called "plasticization" is reflected in an accelerated viscoelastic response.

A very significant portion, of up to 50%, of the water molecules contained within hydrophilic polymers is attributed to molecular interaction with polar sites along the polymeric chains. It was suggested that these sites are concentrated in the vicinity of cross links (Soles and Yee 2000; Musto et al. 2002; Cutugno et al. 2001), thereby relating sorption to cross-link density. The above molecular bonds may be of various kinds, i.e., single, double, or even higher linkages (Cutugno et al. 2001). It thus follows that most epoxies contain two phases of water, namely "bound" and "free." Even so, in the presence of some weak water-polymeric bonds, both bound and free phases are interchangeable and retain mobility. This may be reinforced by the experimental observations that free volume alone does not correlate with diffusivity D (Li et al. 2005).

While there is no direct correlation between V_f and the magnitude of the saturation weight gain M_∞, the latter is highly affected by resin polarity. Nevertheless, for cross-linked polymers, polarity and free volume are inter-related because both are concentrated about cross-linking sites (Soles and Yee 2000).

The above suppositions are supported by weight gain data, that according to theoretical predictions for two phase diffusion should follow curve "A" in Fig. 4.2 rather than Fickian behavior designated by "LF" therein. Such data are given for various polymers (Olmos et al. 2006) as well as by many others.[1]

In hydrophobic polymers it was detected that most water molecules were present in the gaseous phase (Kusanagi and Yukawa 1994). Consequently, in the absence of hydrogen bonds, water migration proceeds entirely by diffusion since no molecular bonds can occur. In this case local scale molecular motions, like those associated with the β transition, play a significant role in the sorption process.

The lack of polar sites enhances the diffusivity D, but energetic considerations suggest that the dependence of D on temperature T is stronger in the presence of polarity (Soles and Yee 2000). It was also possible to employ micro-level arguments to explain the discrepancy between D_a and D_d, where subscripts "a" and "d" refer to absorption and desorption (Soles et al. 2000).

Upon employment of refined instrumentation, it was possible to estimate the average distance between nanopores in a 60% cross-linked rigid resin and thereby the jump length and the jump frequencies of the molecular water random walk process. These turned out to have the approximate value of 12–13 Å and 1.5×10^5–135×10^5 Hz for temperature varying between 5 and 90°C (Soles and Yee 2000).[2]

---

[1] See listing in Table 4.1.

[2] Note that the relationship employed therein is $D = (1/6)a^2\omega$ instead of $(1/2)a^2\omega$ for the one-dimension scheme presented in Sect. 3.2.

It was also noted that at sufficiently high temperatures diffusion and hydrolysis may occur simultaneously (Abeysinghe et al. 1982) and that hydrolysis lowers T_g (Zhou and Lucas 1999). A theoretical model that employs probabilistic considerations for chain scissions by chemically reacted water, and thereby weight loss, had successfully predicted such combined weight gain/weight loss data (Xiao and Shanahan 1997). Those data correspond to curve "D" in Fig. 4.2.

In addition, it was shown by thermal-energetic arguments that M_∞ may increase, decrease, or be independent of temperature T (Li et al., private communication; Medras et al., private communication).

The availability of bonding sites and free volume depends strongly on departures from stoichiometry of various polymeric components, like those in epoxy. Such departures may occur within the interphase regions surrounding the fibers bringing about the accumulation of water therein (Gupta et al. 1985; Grave et al. 1998).

3.2 A Brief Review of Classical Diffusion

Diffusion is a process by which matter is transported from one part of a system to another as a result of random molecular motions (Ghez 1988). A discretized model for isotropic one-dimensional diffusion can be viewed as a random walk model along, say, the x axis. Accordingly, let N_i denote the number of particles located at sites $x_i = ia$, $i = 0, \pm1, \pm2, \ldots$ and assume that each particle can jump to adjacent sites with an equal frequency ω that does not depend on i. Thus, $\frac{1}{2}\omega N_i$ particles jump from site i to site $i + 1$ per unit time, while $\frac{1}{2}\omega N_{i+1}$ particles jump from $i + 1$ to i and similarly between sites $i - 1$ and i. Omitting the details which are derived in the monograph (Ghez 1988), one obtains the following expressions.

$$\frac{\partial N}{\partial t} = \frac{1}{2}\omega a^2 \frac{\partial^2 N}{\partial x^2} + 0(a^4). \tag{3.1}$$

The flux – concentration relations

$$J(x,t) = -\frac{1}{2}\omega a \frac{\partial N}{\partial x}. \tag{3.2}$$

And, upon combining (3.1) and (3.2)

$$\frac{\partial N}{\partial t} = -a \frac{\partial J}{\partial x}. \tag{3.3}$$

Upon assuming, a priori, a continuous mass distribution $m(x,t) = aN$ one obtains the counterparts to expressions (3.1)–(3.3), namely

$$J = -D\frac{\partial m}{\partial x}, \tag{3.4}$$

$$\frac{\partial m}{\partial t} = \frac{-\partial J}{\partial x}, \tag{3.5}$$

$$\frac{\partial m}{\partial t} = D\frac{\partial^2 m}{\partial x^2} \tag{3.6}$$

where $D = (1/2)\omega a^2$ is the diffusivity.

Thus, D is directly proportional to the frequency of mass motion, with dimension of L^2/t, or to mass velocity x length. It thus reflects speed of mass motion, despite the absence of the notion of velocity in parabolic differential equations.

Equations (3.4)–(3.6) are the one-dimensional version of Fick's law where (3.5) accounts for conservation of mass while (3.4) is a constitutive assumption.

To obtain specific solutions to the field (3.6) it is necessary to accompany it with boundary and initial conditions. Such solutions abound in the literature (Crank 1980; Carlslaw and Jaeger 1986; Luikov 1968).

Note that the parabolic (3.6) predicts moisture propagation at an infinite speed. To overcome this unrealistic result, it has been suggested (Taitel 1972) that the above expression be modified by the telegraph equation, namely.

$$\frac{1}{v^2}\frac{\partial^2 m}{\partial t^2} + \frac{1}{D}\frac{\partial m}{\partial t} = \frac{\partial^2 m}{\partial x^2}, \tag{3.7}$$

where v denotes velocity.

Alternately, it has been suggested that all sufficiently small values of moisture content, say below those that fall within data scatter, should be discarded, thus avoiding the infinite velocity paradox.

3.3 The Simplest Classical Solution for Diffusion in a Plate

Consider an initially dry plane sheet (which represents a thin flat coupon), occupying the region $-\ell < x < \ell$, subjected to a uniform ambient fluid concentration m_1 at time $t = 0$.

The solution $m(x,t)$ is given in two forms (Crank 1980):

$$\frac{m(x,t)}{m_1} = 1 - \frac{4}{\pi}\sum_{n=0}^{\infty}\frac{(-1)^n}{2n+1}\exp[-D(2n+1)^2\pi^2 t/4\ell^2]\cos\frac{(2n+1)\pi x}{\ell} \tag{3.8}$$

or

$$\frac{m(x,t)}{m_1} = \sum_{n=1}^{\infty}(-1)^n\left[\mathrm{erfc}\frac{(2n+1)\ell - x}{2\sqrt{Dt}} + \mathrm{erfc}\frac{(2n+v\ell + x}{2\sqrt{Dt}}\right]. \tag{3.9}$$

The total weight gain, namely

$$M(t) = \int_{-\ell}^{\ell} m(x,t)\,\mathrm{d}x, \tag{3.10}$$

is also given in two corresponding forms. Thus,

$$M(t)/M_\infty = 1 - \sum_{n=1}^{\infty}\frac{8}{(2n+1)^2\pi^2}\exp[-\pi^2(2n+1)^2t^*] \tag{3.11}$$

or

$$M(t)/M_\infty = 4\sqrt{t^*}\left[\pi^{-1/2} + \sum_{n-1}^{\infty}(-1)^n i\,\mathrm{erfc}\frac{n}{\sqrt{t^*}}\right] \tag{3.12}$$

in the above $t^* = Dt/(2\ell)^2$, erfc(·) denotes the complementary error function, $i\,\mathrm{erfc}(z) = \int_z^\infty \mathrm{erfc}(u)\,\mathrm{d}u$, and M_∞ is the weight gain at saturation, i.e., $M_\infty = 2\ell m_1$.

The expression (3.11) converges rapidly for large values of t^*, (in fact, accounting for data scatter, one term would definitely suffice for $t^* > 0.5$), while (3.12) converges rapidly for small values of t^* (one term would definitely suffice for $t^* < 0.5$). More specific details are given elsewhere (Weitsman 1981).

Note that the non-dimensional time t^* varies like the square of the thickness 2ℓ, thus so does the time-to-saturation.

A plot of $M(t^*)/M_\infty$ vs. $\sqrt{t^*}$ is given in Fig. 3.1. As may be seen, the above ratio is given by a straight line, of slope $4/\sqrt{\pi}$, for $0 \le t^* \le 0.557$ (i.e., by the leading term in (3.12)) with $M(t^*)/M_\infty = 0.62$ and saturation is attained, to within 1% error, at $t^* = 1.32$.

The above observation enables the early experimental evaluation of the diffusion coefficient D, as follows:

Take intermittent weight gain measurements on a flat sample of thickness 2ℓ, at approximately equal $\sqrt{\text{time}}$ intervals, up to such times when data depart from a straight line. The point of departure corresponds to $D \sim 0.557\ (2\ell)^2/t$.

Thus far, the diffusion process has been entirely analogous to that of heat conduction in solids, with the exception that for both polymers and polymeric composites D is of order 10^{-7} mm^2/s which is about 6 orders of magnitude smaller than the thermal diffusivity k.

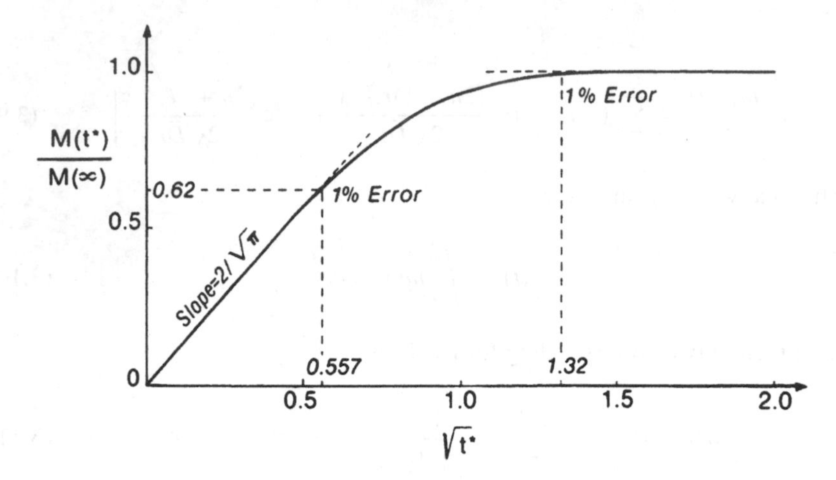

Fig. 3.1 The total weight gain $M(t)/M(\infty)$ vs. $\sqrt{t^*}$according to Fick's law, with locations where departure from straight lines exceed 1%

3.4 Interfacial Conditions

Consider the case of two distinct polymeric, or polymeric composite, plates with different diffusivities D_1, and D_2, as well as saturation levels, $M_\infty{}^{(1)}$and $M_\infty{}^{(2)}$, in contact at, say $x = 0$.

Continuity of flux, i.e., conservation of mass, at $x = 0$ requires

$$D_1 \frac{\partial m_1}{\partial x} = D_2 \frac{\partial m_2}{\partial x}. \tag{3.13}$$

However, continuity of the chemical potentials, i.e., $\mu_1 = \mu_2$, precludes the continuity of m_1 and m_2 at $x = 0$.

Starting with the basic equations for an ideal monatomic gas and extending them to the case of a multi-component ideal system (Haase 1969) one has

$$\mu_i = \hat{\mu}_i(p, T) + RT\ell n(m_i). \tag{3.14}$$

For real systems it is necessary to replace m_i by the "activity" a_i, which may take the form of $a_i = v_i m_i$ (no sum on i) (Haase 1969).

The requirement that $\mu_1 = \mu_2$ at $x = 0$ translates into

$$m_2 = \hat{\alpha} m_1, \tag{3.15}$$

where $\hat{\alpha} = (v_2/v_1)\exp[(\hat{\mu}_1 - \hat{\mu}_2)/RT]$.

At equilibrium $m_1 = M_1(\infty)/L_1$ and $m_2 = M_2(\infty)/L_2$, thus, for diffusion processes that exhibit clear equilibrium weight gain values one has

$$\hat{\alpha} = [M_2(\infty)/M_1(\infty)](L_1/L_2). \tag{3.16}$$

Equations (3.13) and (3.15) are the interfacial conditions at $x = 0$, subject to (3.16).

Expressions (3.13)–(3.16) are readily extended to the case of n layers (Clark 1983).

Consider for example a symmetric, initially dry, bi-material lay-up exposed to a constant ambient environment. Let the interior region, with concentration m_1, and diffusivity D_1, occupy the region $-a < x < a$, and similarly let m_2 and D_2 correspond to the outer regions $a < |x| < L/2$. Similarly, let $m_1(\infty) = M_1(\infty)/2a$ and $m_2(\infty) = M_2(\infty)/(L - 2a)$ denote the equilibrium concentration levels of each material. Therefore, $\hat{\alpha} = m_2(\infty)/m_1(\infty)$.

We thus have

$$\frac{\partial m_1}{\partial t} = D_1 \frac{\partial^2 m_1}{\partial x^2} \tag{3.17a}$$

and

$$\frac{\partial m_2}{\partial t} = D_2 \frac{\partial^2 m_2}{\partial x^2} \tag{3.17b}$$

with initial and boundary conditions such as $m_0(x,0) = 0$, and $m(\pm L/2,t)$ (=constant) and interfacial condition $x = \pm a$

$$m_2(\pm a, t) = \hat{\alpha} m_1(\pm a, t), \tag{3.18}$$

$$D_1 \frac{\partial m_1}{\partial x}(\pm a, t) = D_2 \frac{\partial m_2}{\partial x}(\pm a, t). \tag{3.19}$$

In the above equations, m_1 denotes the moisture content in the central layer $-a < x < a$ and m_2 corresponds to the outer layers $a < |x| < L/2$.

The solution to (3.17a)–(3.19) together with the aforementioned initial and boundary conditions can be obtained through the method of homogenization (Wirth and Rodin 1982). Accordingly, we let $m_2^* = m_2 - m^b$, $m_1^* = m_1 - m^b/\hat{\alpha}$, whereby the boundary value problem reduces to

$$\frac{\partial m_1^*}{\partial t} = D_1 \frac{\partial^2 m_1^*}{\partial x^2}, \tag{3.20a}$$

$$\frac{\partial m_2^*}{\partial t} = D_2 \frac{\partial^2 m_2^*}{\partial x^2} \tag{3.20b}$$

and

$$m_2^*\left(\pm\frac{L}{2}, t\right) = 0, \tag{3.21a}$$

$$m_2^*(\pm a, t) = \hat{\alpha} m_1^*(\pm a, t), \tag{3.21b}$$

$$D_1 \frac{\partial m_1^*}{\partial x}(\pm a, t) = D_2 \frac{\partial m_2^*}{\partial x}(\pm a, t) \tag{3.21c}$$

with the non-zero initial conditions

$$\begin{aligned} m_1^*(x,0) &= -m^b/\alpha, \\ m_2^*(x,0) &= -m^b. \end{aligned} \tag{3.22}$$

Employing separation of variables $m^*(x,t) = X(x)T(t)$, together with the symmetry condition $\partial \mu_1(0,t)/\partial x = 0$ we obtain

$$m_\ell^*(x,t) = \sum_{k=1}^{\infty} X_{\ell k}(x) \exp(-\lambda_k^2 t) \quad \ell = 1, 2, \tag{3.23}$$

where

$$X_{1k}(x) = C_{1k} \cos\left(\frac{\lambda_k x}{D_1^{1/2}}\right),$$

$$X_{2k} = B_{2k} \sin\left(\frac{\lambda_k x}{D_2^{1/2}}\right) + C_{2k} \cos\left(\frac{\lambda_k x}{D_2^{1/2}}\right).$$

The three conditions listed in (3.21) yield a determinantal equation for the eigenvalues λ_k as follows

$$\tan\left(\frac{\lambda_k a}{D_1^{1/2}}\right) = \hat{\alpha}\left(\frac{D_1}{D_2}\right)^{1/2} \cot\left(\lambda_k \frac{L/2 - a}{D_2^{1/2}}\right). \tag{3.24}$$

The roots λ_k in (3.24) must be determined numerically, where the ratios B_{2k}/C_{1k} and C_{2k}/C_{1k} can be obtained for each value of k, and without loss of generality we can choose $C_{1k} = 1$

With the latter choice, we have

$$B_{2k} = -\left(\frac{D_1}{D_2}\right)^{1/2} \sin\left(\frac{\lambda_k a}{D_1^{1/2}}\right)\left\{\cos\left(\frac{\lambda_k a}{D_2^{1/2}}\right)\left[1 + \tan\left(\frac{\lambda_k L}{2D_2^{1/2}}\right)\tan\left(\frac{\lambda_k a}{D_2^{1/2}}\right)\right]\right\}^{-1},$$

$$C_{2k} = -B_{2k}\tan\left(\frac{\lambda_k L}{2D_2^{1/2}}\right).$$

Further manipulations finally yield (Clark 1983)

$$m_1^*(x,t) = \sum_{k=1}^{\infty} F_k X_{1k}(x)\exp(-\lambda_k^2 t), \tag{3.25}$$

where $F_k = N_k/D_k$ with

$$N_k = -\frac{m^b}{D_1\hat{\alpha}\lambda_k}\left\{c_{1k}D_1^{1/2}\sin\left(\frac{\lambda_k a}{D_1^{1/2}}\right) + C_{2k}D_2^{1/2}\left[\sin\left(\frac{\lambda_2 b}{D_2^{1/2}}\right) - \left(\frac{\lambda_k a}{D_2^{1/2}}\right)\right]\right.$$
$$\left. - B_{2k}D_2^{1/2}\left[\cos\left(\frac{\lambda_k b}{D_2^{1/2}}\right) - \cos\left(\frac{\lambda_k a}{D_2^{1/2}}\right)\right]\right\},$$

$$D_k = \frac{1}{D_1\hat{\alpha}}\left(C_{2k}^2\left\{\frac{b-a}{2} + \frac{D_2^{1/2}}{4\lambda_k}\left[\sin\left(\frac{2\lambda_k b}{D_2^{1/2}}\right) - \sin\left(\frac{2\lambda_k a}{D_2^{1/2}}\right)\right]\right\}\right.$$
$$+ D_2^{1/2}\frac{B_{2k}C_{2k}}{2\lambda_k}\left[\cos\left(\frac{2\lambda_k b}{D_2^{1/2}}\right) - \cos\left(\frac{2\lambda_k a}{D_2^{1/2}}\right)\right] + B_{2k}^2\left\{\frac{b-a}{2} - \frac{D_2^{1/2}}{4\lambda_k}\left[\sin\left(\frac{2\lambda_k b}{D_2^{1/2}}\right) - \sin\left(\frac{2\lambda_k a}{D_2^{1/2}}\right)\right]\right\}$$
$$+ \frac{1}{D_1}\left[\frac{a}{2} + \frac{D_1^{1/2}}{4\lambda_k}\sin\left(\frac{2\lambda_k a}{D_1^{1/2}}\right)\right]. \tag{3.26}$$

The actual moisture distributions $m_1(x,t)$ and $m_2(x,t)$ are then obtainable from $m_1(x,t) = m_1^*(x,t) + m^b/\hat{\alpha}$ and $m_2(x,t) = m_2^*(x,t) + m^b$. The total moisture content is determined by integration across the laminate thickness.

A similar solution was subsequently developed for hybrid IM7/BMI composites employing a numerical method (Tang et al. 2005).

3.5 Concentration-Dependent Diffusion

This circumstance is related by a concentration-dependent diffusion coefficient, namely $D = D(m)$.

In this case, (3.6) is modified to read

$$\frac{\partial m}{\partial t} = \frac{\partial}{\partial x}\left[D(m)\frac{\partial m}{\partial x}\right]. \tag{3.27}$$

Note that in this case coupling involves the entire history of m because it varies with both x and t. The characterization of $D = D(m)$ is a formidable challenge, though some insights may be gained by collecting absorption and desorption weight gain and weight loss data on samples with several distinct thicknesses.

3.6 Mobile and Bound Penetrant Molecules: Two Phase Diffusion

In some cases, a portion of the diffusing molecules may become attached to or trapped within some sites inside the polymeric chains. Such attachment may be temporary and an exchange may occur over time between mobile and bound molecules. This phenomenon is denoted as a "Two-Stage Diffusion" and, in the one-dimensional case, is formulated as shown below (Carter and Kibler 1978; Gurtin and Yatomi 1979)

Let m_{m} and m_{b} refer to mobile and bound fluid content, respectively, and let γ and β denote the rates at which m_{b} becomes mobile and m_{m} becomes bound, respectively. Also, let D denote the diffusion coefficient of the mobile portion. If follows that, upon viewing m_{b} a source as term, one has

$$D\frac{\partial^2 m_{\mathrm{m}}}{\partial x^2} = \frac{\partial(m_{\mathrm{m}} + m_{\mathrm{b}})}{\partial t} \tag{3.28a}$$

and

$$\frac{\partial m_{\mathrm{b}}}{\partial t} = \gamma m_{\mathrm{m}} - \beta m_{\mathrm{b}}. \tag{3.28b}$$

The one-dimensional solution for an initially dry sample of thickness h, upon assuming that both $\beta, \gamma \ll k = \pi^2 D/h^2$ reads (Carter and Kibler 1978).[3]

[3] The general solution was expressed by Gurtin and Yatomi (1979). An illuminating discussion on the subject is given in a review article by Bond and Smith (2006).

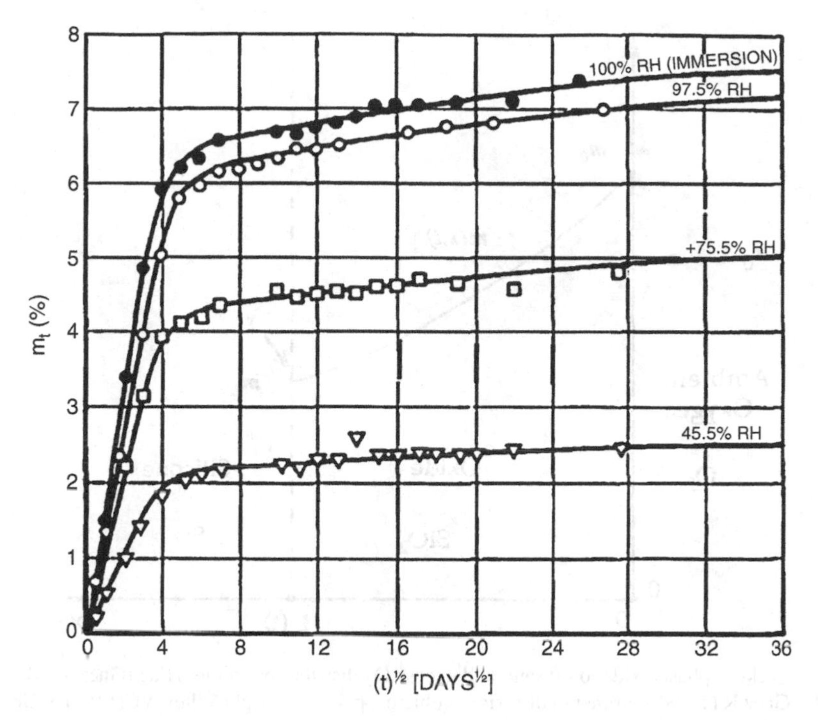

Fig. 3.2 Moisture uptake in 5,208 resin at 24°C (75°F). Solid curves from (3.29), using same values of parameters for each humidity (Carter and Kibler 1978)
Credit: Carter HG, Kibler KG "Langmuir-type model for anomalous moisture diffusion in composite resins." 12(2):118–131, copyright 1978 by *Journal of Composite Materials*, Reprinted by permission of SAGE

$$m_t \simeq m_\infty \left\{ \frac{\beta}{\gamma + \beta^{e-\gamma t}} \left[1 - \frac{8}{\pi^2} \sum_{l=1(\text{odd})}^{\infty} \frac{e^{-kl^2 t}}{l^2} \right] + \frac{\beta}{\gamma + \beta}(e - \beta t - e^{-\gamma t}) + (1 - e^{-\beta t}) \right\}. \tag{3.29}$$

Typical weight gain data for two phase diffusion are shown in Fig. 3.2.

3.7 Diffusion Associated with Chemical Reaction

In several circumstances when a solid is exposed to a specific harsh environment a chemical reaction takes place that transforms the atomic or molecular structure of the solid and splits it into two phases. Such events may occur within polymers exposed to acids, glass fibers exposed to water, silicon exposed to oxygen, etc.

In this case, the reaction-modified phase arises within the region containing the diffusing molecules, while the unreacted phase remains outside the region where diffusion takes place. These distinct phases are separated by a moving reaction front.

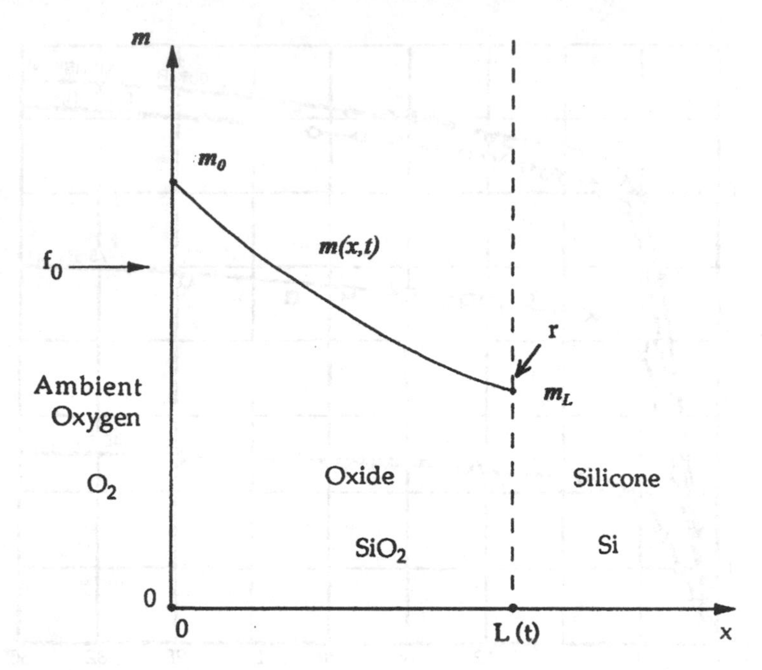

Fig. 3.3 Three-phase oxidation system and excess O_2 distribution in the oxide (Ghez 1988) Credit: Ghez R (1988) A primer of diffusion problems, p 48. Copyright Wiley-VCH Verlag GmbH & Co. KGaA. Reproduced with permission

In the one-dimensional case, this process is described by Fig. 3.3 sketched above (Ghez 1988; Cussler 2009).

Flux balance at the location $L(t)$ of the reaction front is given by

$$D\frac{\partial m}{\partial x}\bigg|_{x=L} + r + m_L\frac{\mathrm{d}L}{\mathrm{d}t} = 0, \tag{3.30}$$

where r is the reaction rate.

3.8 Effects of Fluid's Chemistry

Fluid chemistry and molecular structure have a significant effect on the sorption process. This is demonstrated by observations that the saturation levels of distilled water, tap water and sea water seem to decrease in the above order while the increasing activity levels of water, methylene chloride (CH_2Cl_2) and n-hepatine correspond to higher saturation levels in PEEK polymer. In contrast with water and methylene chloride the latter chemical exhibits weight-gain data akin to that of a two-stage diffusion process. Similarly, increasing levels of fluid's acidity have strong effects on the sorption process, while the ingress of high molecular weight fluids, like oils, maybe substantially impeded.

It was also recorded that certain chemicals caused scissions within polymeric chains and of side groups. Relevant data are given in Table 4.1, demonstrating a strong dependence on temperature and activity that brings about weight losses in exposed samples.

3.9 Coupling Between Diffusion and Polymeric Relaxation: Two-Stage Diffusion

This combined effect is dominated by two distinct yet inter-related time-dependent phenomena, diffusion across the thickness of, say, a polymeric slab and time-dependent relaxation of the polymeric chains. Since saturation time, t_s, is thickness (h) dependent (roughly $t_s \sim h^2$), and relaxation time is enhanced by fluid content (through a "reduced time" $\xi(t) = t/a_H(m)$), it is obvious that the character of the coupling process will vary in essential details for polymeric slabs of distinct thicknesses. In the extreme case of thin films, the diffusion process will attain equilibrium within a very short time, resulting in the establishment of a uniform distribution across the film much sooner than the time required for the evolution of the relaxation process, thereby the process of fluid ingress is dominated by the diffusivity of an unrelaxed polymer. At the other extreme, namely that of a very thick slab, the diffusion process is slower than that of relaxation and fluid ingress tends to be governed by the diffusivity of fully relaxed polymeric chains. For intermediate slab thicknesses the process is fully coupled and will depend on the entire history of fluid ingress process. Since a_H is $a_H(m)$ and $m = m(x, t)$ it is clear that this dependence will also vary with location x, thus presenting a classical case of a "moving target." Typical weight gain data, exhibiting earlier results for an unrelaxed polymer, followed by the combined effects of diffusion and relaxation, are shown in Fig. 3.4. A detailed discussion of this topic is presented in Chap. 6.

3.10 Coupling Between Fluid Ingress and Mechanical Strain or Stress

Since the configurations of the polymeric chains are affected by mechanics-based factors, as well as by the magnitude of the free volume, they necessarily interact with the diffusion process. Unlike the aforementioned case of creep, the imposition of a constant strain would presumably maintain all fully relaxed polymeric chains at nearly fixed configurations during fluid ingress and thus yield data that may be largely uncoupled. Accordingly to this supposition such data – collected

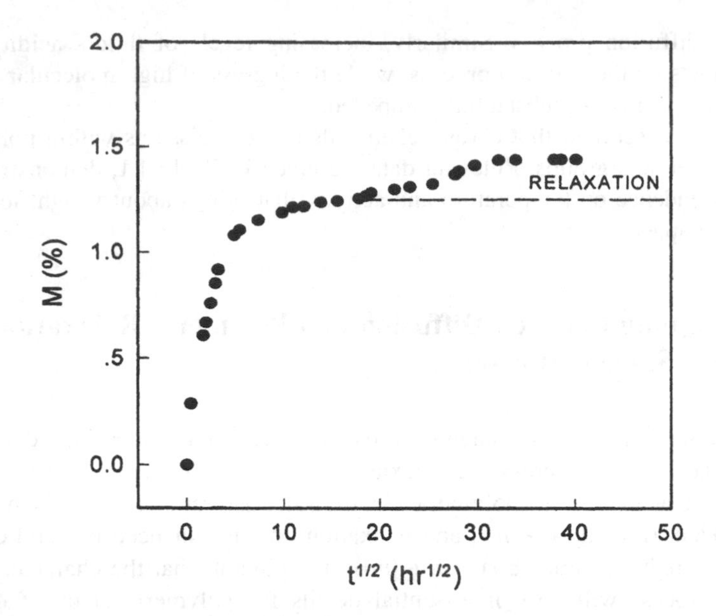

Fig. 3.4 Additional water uptake of unidirectional laminate measured at 70°C exhibiting two-equilibrium stages seemingly due to polymer swelling and physical relaxation (Suh et al. 2001) Credit: Suh D, Ku M, Nam J, Kim B, Yoon S. "Equilibrium water uptake of epoxy/carbon fiber composites in hygrothermal environmental conditions." 35(3):264–278, copyright 2001 by *Journal of Composite Materials*. Reprinted by Permission of SAGE

at several distinct levels of constant strain – may simplify the quantification of the coupled effects due to time-dependent molecular motion and time-dependent diffusion. In contrast, the application of constant stress will induce time-dependent deformation (i.e., creep), thus resulting in a coupling of two time-dependent phenomena (creep and diffusion). A detailed model is given in Sect. 6.3.

3.11 Coupling Between Diffusion and Damage

It was noted that fluid-induced swelling would induce non-uniform hygral stresses in the regions occupied by the polymeric phase in fiber-reinforced composites. This non-uniformity stems from the inherent geometry of the fiber-matrix cross section and is enhanced by the random distribution of fibers positioned therein, as shown in Fig. 3.5 (Daniel and Ishai 2005)

The aforementioned stress concentrations, some of which are tensile, will induce micro-cracks within the composite that increase its fluid absorption capacity. Several models on this topic are presented in Sect. 5.7.

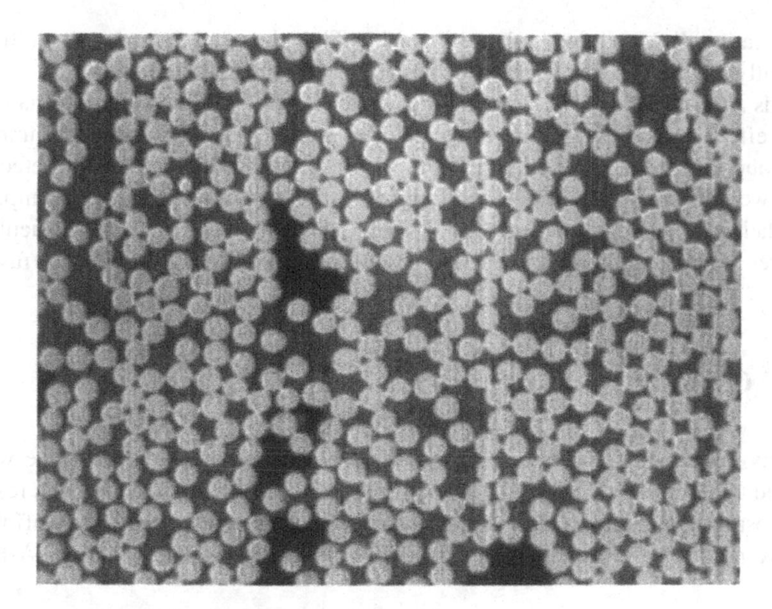

Fig. 3.5 Digitize image of carbon/epoxy cross section used for determination of fiber volume and void volume ratios (Daniel and Ishai 2005)

3.12 Fluid Ingress Within Fibers

Both aramid (Kevlar) and glass fibers absorb fluids. Consequently, the overall ingress of fluids is affected by the sorption within two disparate phases, namely the fibers and the polymeric resin. Typically, the combined effect may resemble the weight gain data representative of a two-stage diffusion.

3.13 Capillary Motion

Capillary paths arise within fiber-reinforced composites due to mechanically or hygro-thermally induced micro cracks that are channeled parallel to the reinforcement directions. Such cracks can occur under both static and cyclic mechanical loads, and may also develop under sequential thermal "spiking." Exposure to ambient environment will instigate the flow of fluid into those capillaries. Specific details will be given in Sect. 4.10.

3.14 Wicking Action

This process of fluid ingress occurs in fiber-reinforced composites along the fiber/matrix interphase regions. These regions, which surround the fibers and are typically a fraction of micron in thickness, consist of a boundary layer of configurationally modified polymeric resin. The chemical interphase mixtures, especially if they diverge from stoichiometry, tend to absorb excessive amounts of fluids and introduce paths for wicking action. It may be possible to estimate the above effect by exposing composite material samples to ambient environments, with coupons cut perpendicularly to the fiber direction, as well as by detecting larger weight gains for samples with higher ratios of edge surfaces as compared with their top and bottom surfaces. Though data vary and may be difficult to interpret, it appears that wicking action proceeds at a 10 times faster than diffusion.

3.15 Osmosis

In many circumstances polymers may act as a semi-permeable membrane when exposed to a mixture of external vapor or fluid molecules. This phenomenon results in an osmotic pressure that induces internal damage and affects the diffusion process, rendering it sensitive to the detailed composition of such mixture (Ashbee 1989).

3.16 Relative Humidity (RH) and Equilibrium Moisture Content (M_∞)

For many vapors and polymeric composites the aforementioned quantities were related empirically by the expression

$$M_\infty = a_\mathrm{m}(\mathrm{RH}/100)^{b_\mathrm{m}}, \quad (3.31)$$

Where a_m and b_m are constants, and RH is in %.
Other forms are also possible.

3.17 Temperature Dependence of Diffusion Coefficient

The diffusion coefficient D, which affects the time-to-saturation, is extremely sensitive to temperature T. Typically, this temperature dependence is expressed by an Arrhenius type equation

$$D(T) = Ae^{-B/T} \quad (T \text{ in °K}), \quad (3.32)$$

Where A and B are empirical constants that may depend weakly on the fluid content m.

For some materials it is impossible to express $D(T)$ by (3.32) for a sufficient range of temperatures, and two values of A and of B are necessary for disparate spans of T.

3.18 Time Scales

1. The moisture diffusion coefficient D is of order $10^{-7}\text{mm}^2/\text{s}$, which for composites and polymers is about 6 and 7 orders of magnitudes smaller, respectively, than the temperature diffusion coefficient. The above processes can thus be decoupled.
2. Capillary motion occurs at about 6 orders of magnitude faster than diffusion. Both processes can be decoupled, with consequent ramifications for diffusion and damage.
3. Wicking along the fiber/matrix interphase occurs about 10 times faster than diffusion within the polymeric phase. The two phenomena are weakly coupled.
4. Diffusion and polymeric creep occur on approximately similar time-scales. Except for the limiting cases of very thin or very thick plate thicknesses these phonomena are inherently coupled.

References

Abeysinghe H, Edwards W, Pritchard G, Swampillai G (1982) Degradation of crosslinked resins in water and electrolyte solutions. Polymer 23(12):1785–1790

Ashbee KH (1989) Fundamental principles of fiber reinforced composites. CRC Press, Boca Raton

Bond DA, Smith PA (2006) Modeling the transport of low-molecular-weight penetrants within polymer matrix composites. Appl Mech Rev 59(5):249–268

Carlslaw HS, Jaeger JC (1986) Conduction of heat in solids. Oxford University Press, Oxford

Carter HG, Kibler KG (1978) Langmuir-type model for anomalous moisture diffusion in composite resins. J Compos Mater 12(2):118–131

Clark DL (1983) Moisture absorption in hybrid composites. Texas A&M University Report MM 4665-83-16, December

Crank J (1980) The mathematics of diffusion, 2nd edn. Oxford University Press, Oxford

Cussler EL (2009) Diffusion: mass transfer in fluid systems. Cambridge University Press, Cambridge

Cutugno S, Martelli G, Negro L, Savoia D (2001) The reaction of β-amino alcohols with 1,1′-carbonyldiimidazole – influence of the nitrogen substituent on the reaction course. Eur J Org Chem 2001(3):517–522

Daniel IM, Ishai O (2005) Engineering mechanics of composite materials. Oxford University Press, New York

Deng Q, Jean YC (1993) Free-volume distributions of an epoxy polymer probed by positron annihilation: pressure dependence. Macromolecules 26(1):30–34

Ghez R (1988) A primer of diffusion problems. Wiley-Interscience, New York

Grave C, McEwan I, Pethrick RA (1998) Influence of stoichiometric ratio on water absorption in epoxy resins. J Appl Polym Sci 69(12):2369–2376

Gupta VB, Drzal LT, Rich MJ (1985) The physical basis of moisture transport in a cured epoxy resin system. J Appl Polym Sci 23(4–6):4467–4493

Gurtin ME, Yatomi C (1979) On a model for two phase diffusion in composite materials. J Compos Mater 13(2):126–130

Haase R (1969) Thermodynamics of irreversible processes. Addison-Wesley, New York (especially section 1–19)

Kawagoe M, Hashimoto S, Nomiya M, Morita M, Qiu J, Mizuno W, Kitano H (1999) Effect of water absorption and desorption on the interfacial degradation in a model composite of an aramid fibre and unsaturated polyester evaluated by Raman and FT infra-red microspectroscopy. J Raman Spectrosc 30(10):913–918

Kusanagi H, Yukawa S (1994) Fourier transform infra-red spectroscopic studies of water molecules sorbed in solid polymers. Polymer 35(26):5637–5640

Li L, Zhang S, Chen Y, Liu M, Ding Y, Luo X, Pu Z, Zhou W, Li S (2005) Water transportation in epoxy resin. Chem Mater 17(4):839–845

Luikov AV (1968) Analytical heat diffusion theory. Academic, New York

Marechal Y, Chamel A (1996) Water in a biomembrane by infrared spectrometry. J Phys Chem 100(20):8551–8555

Musto P, Ragosta G, Scarinzi G, Mascia L (2002) Probing the molecular interactions in the diffusion of water through epoxy and epoxy-bismaleimide networks. J Polym Sci B Polym Phys 40(10):922–938

Olmos D, López-Morón R, González-Benito J (2006) The nature of the glass fibre surface and its effect in the water absorption of glass fibre/epoxy composites. The use of fluorescence to obtain information at the interface. Compos Sci Technol 66(15):2758–2768

Sammon C, Mura C, Yarwood J, Everall N, Swart R, Hodge D (1998) FTIR – ATR studies of the structure and dynamics of water molecules in polymeric matrixes. A comparison of PET and PVC. J Phys Chem B 102(18):3402–3411

Soles CL, Yee AF (2000) A discussion of the molecular mechanisms of moisture transport in epoxy resins. J Polym Sci B Polym Phys 38(5):792–802

Soles CL, Chang FT, Gidley DW, Yee AF (2000) Contributions of the nanovoid structure to the kinetics of moisture transport in epoxy resins. J Polym Sci B Polym Phys 38(5):776–791

Suh D, Ku M, Nam J, Kim B, Yoon S (2001) Equilibrium water uptake of epoxy/carbon fiber composites in hygrothermal environmental conditions. J Compos Mater 35(3):264–278

Taitel Y (1972) On the parabolic, hyperbolic and discrete formulation of the heat conduction equation. Int J Heat Mass Transf 15(2):369–371

Tang X, Whitcomb JD, Li Y, Sue H (2005) Micromechanics modeling of moisture diffusion in woven composites. Compos Sci Technol 65(6):817–826

Weitsman Y (1981) A rapidly convergent scheme to compute moisture profiles in composite materials under fluctuating ambient conditions. J Compos Mater 15:349–358

Wirth P, Rodin E (1982) A unified theory of linear diffusion in laminated media. Adv Heat Transf 15:284–321

Xiao GZ, Shanahan MER (1997) Water absorption and desorption in an epoxy resin with degradation. J Polym Sci B Polym Phys 35(16):2659–2670

Zhou J, Lucas JP (1999) Hygrothermal effects of epoxy resin. Part I: the nature of water in epoxy. Polymer 40(20):5505–5512

Chapter 4
Fluid Sorption Data and Modeling

4.1 Types of Weight-Gain Data

A qualitative correlation between temperature T, penetrant activity a, and the forms of fluid penetration into polymers is given in the map depicted in Fig. 4.1 (Hopfenberg and Frisch 1969).

Upon reviewing a large amount of data, it became possible to relate fluid sorption in polymers and polymeric composites by means of six schematic curves relating weight gain to $\sqrt{t^*}$ as sketched in Fig. 4.2 below. The scales marked in that figure apply to the linear Fickian (LF) plot alone.

Curves "A" and "B" accord with data collected for both polymers and composites. Of these, "A" corresponds to a case where weight-gain never attains equilibrium, such as for two phase diffusion, and "B' represents the circumstance of two-stage diffusion. Weight gain data compatible with those curves are associated with benign fluid effects that are essentially reversible upon drying.

Data represented by curves "C" and "D" are recorded mostly for polymeric composites. Of these, "C" accounts for the case of rapidly increasing fluid content, which is usually accompanied by damage growth that leads to material break down, large deformations, as well as occasional failure.

Curve "D" accords with weight loss that is attributable to chemical or physical break-down of the material. This break down occurs mostly in the form of separations between fibers and polymer, caused by leaching along the fiber/matrix interphase regions, as well as by hydrolysis, i.e. the detachment of side groups from the backbones of the polymeric chains followed by chain scissions.

Circumstances that accord with weight gain data along curves "C" and "D" represent irreversible response associated with loss of material integrity, portending possible structural failure.

Curves "S" is at times associated with a moving diffusion front (Nicolais et al. 1991).

It is shown in Table 4.1 below that weight gain data for most polymeric composites are apt to switch in form between the curves "LF", "A–D" in Fig. 4.2.

Y.J. Weitsman, *Fluid Effects in Polymers and Polymeric Composites*,
Mechanical Engineering Series, DOI 10.1007/978-1-4614-1059-1_4,

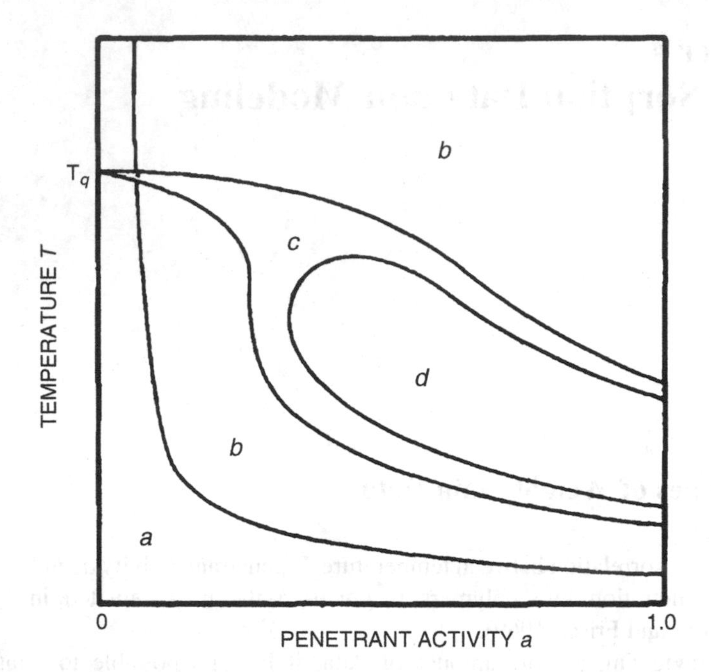

Fig. 4.1 The effect of temperature and activity on the likely sorption kinetic processes for a polymer-penetrant system (**a**) Fickian diffusion, (**b**) concentration dependent diffusion, (**c**) combined relaxation and diffusion (**d**) relaxation controlled diffusion (After Hopfenberg and Frisch 1969) Copyright [Hopfenberg, H. B., and Frisch, H. L. (1969). "Transport of organic micromolecules in amorphous polymers." *Journal of Polymer Science Part B: Polymer Letters*, 7(6), pp. 405–409.] "This material is reproduced with permission of John Wiley & Sons, Inc."

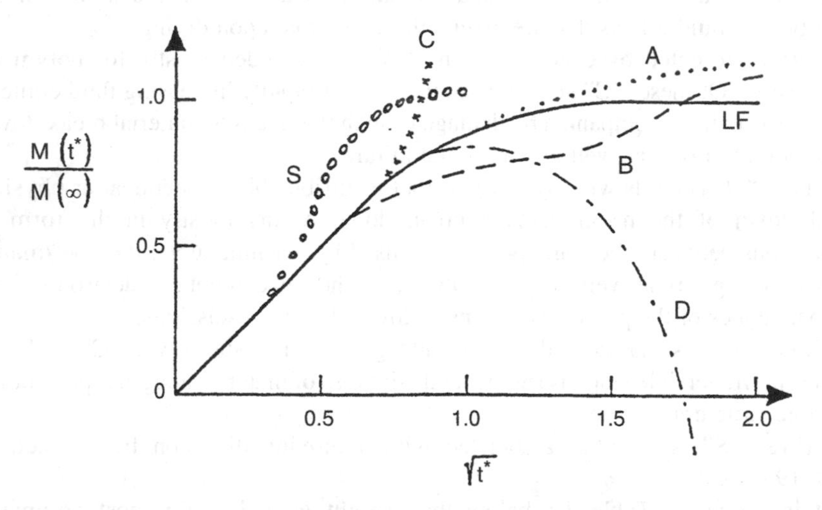

Fig. 4.2 Schematic curves representing four categories of recorded non-fickian weight-gain sorption. The *solid line*, designated by LF, corresponds to linear Fickian diffusion

Table 4.1 Types of sorption data for various materials and exposure

Material system	Exposure	Type of weight-gain data sketched in Fig. 4.2	Reference
T300/5208 gr/ep	T <60 C	LF	Shirrell (1978)
	T >60 C	A	
Epoxy and AS/3501-5 gr/cp	Distilled water at 22–70°C; Humid air water immersion	A and/or B	Whitney and Browning (1978)
T300/1034	Distilled water at 90°C	C	Mohlin (1984)
	98% RH at 70°C	LF or A	
	98% RH at 90°C	A inconclusive	
Glass/polyester	Distilled water at 40°C (no external load)	B	Lagrange et al. (1991)
	Sea water at 40°C various types of polyester	LF or A	
	Distilled water at 40°C and $\sigma = 0.5\sigma_{ult}$	D	
SMC-25	RH = 100%, 32 and 50°C	LF	Loos and Springer (1981)
	RH = 40%, 65°C	LF	
	RH = 60%, 65°C	A	
	RH = 100%, 65°C	D	
	Distilled water, 23°C	A	
	Distilled water, 50°C	D	
	Salt water, 23°C	LF	
	Salt water, 50°C	B	
	Diesel or jet fuel, 23 and 50°C	A	
	Aviation oil, 23°C	LF	
SMC-65	RH = 100%, 32°C	A	Loos and Springer (1981)
	RH = 100%, 50 and 65°C	D	
	RH = 40%, 65°C	LF	
	RH = 60%, 65°C	A	
	RH = 100%, 65°C	D	
	Distilled water, 23°C	A	
	Distilled water, 50°C	D	

(continued)

Table 4.1 (continued)

Material system	Exposure	Type of weight-gain data sketched in Fig. 4.2	Reference
	Salt water, 23°C	LF	
	Salt water, 50°C	A	
	Diesel Fuel, 23°C	A[a]	
	Diesel Fuel, 50°C	LF[a]	
	Jet fuel, 23°C	A[a]	
	Jet fuel, 50°C	LF[a]	
	Aviation oil, 23°C	LF[a]	
	Aviation oil 50°C	LF[a]	
SMC 30EA	RH = 100%, 32°C	A	Loos and Springer (1981)
	RH = 100%, 50 and 65°C	D	
	RH = 40%, 65°C	LF	
	RH = 60%, 65°C	D	
	RH = 100%, 65°C	D	
	Distilled water, 23°C	A	
	Distilled water, 50°C	D	
	Salt water, 23°C	LF	
	Salt water, 50°C	A	
	Diesel fuel, 23°C	A	
	Diesel fuel, 50°C	A	
	Jet fuel, 23°C	A	
	Jet fuel, 50°C	B	
	Aviation Oil, 23°C	LF or A	
	Aviation Oil, 50°C	A	
Glass/epoxy *Diamine hardener*	12–100% RH, 25–90°C	LF	Bonniau and Bunsell (1984)
Glass/epoxy	0–100% RH, 25°C < T <90°C	LF or 2-stage diffusion	
	Water immersion	A or C	
Glass/epoxy	28–100% RH, 25–90°C	A	Springer (1981)

Dicyandiamide hardner			
Glass/epoxy	52–100% RH, 90°C	D	
Anhidridehardener			
XMC-3 adhesive	50% RH, water	LF	Springer (1981)
	Salt water at 23°C	LF	
	Water at at 93°C	D	
T300/914	70–98% RH at 50–70°C	B	Blikstad et al. (1988)
Cured or undercured epoxy	"Low temperatures"	LF	Gupta et al. (1985)
	"Intermediate temperature"	A	
	"High temperatures"	C	
Glass/epoxy	Water immersion at 80°C	D	Hamada et al. (1991)
Glass/vinylester	Water immersion at 40, 60, 80°C	LF	Morii et al. (1991)
Acryl-silane surface treatment	Water immersion at 40, 60°C	LF	
	Water immersion at 90°C	A	
S glass/epoxy	Distilled water at 30–80°C	A	Chateauminois et al. (1991)
	Distilled water at 90°C	C	
E glass/epoxy	Water at 22°C	LF	Dewimille et al. (1980)
	Water at 80–100°C	D	
Epoxy	Water at 22–100°C	A	
Glass/epoxy	Water at 80°C	D	Fukuda (1986)
Hybrid glass/carbon epoxy		D	
Vinyl ester	Water	A	Hojo et al. 1982
	Salt water	D	
S2 glass/epoxy	75% RH at 75°C – first exposure	A	Lo et al. 1982
S2 glass/epoxy	75% RH at 75°C – second and third exposure	B	
Swirl mat glass/PPS	Hot water	D	Lou and Murtha (1988)
gr/PPS	Nonimpacted and preimpacted exposed to jet fuel	A	Ma et al. 1991
APC-2		LF	
CF/epoxy	Sea water	A	Manocha et al. 1982
Hybrid CF/kevlar-epoxy	Sea water	A	

(continued)

Table 4.1 (continued)

Material system	Exposure	Type of weight-gain data sketched in Fig. 4.2	Reference
gl/ep-gr/ep hybrid	Water immersion 70°C	D	French and Pritchard 1991
Glass/UP	Distilled water at 23°C	LF	Menges and Gitschner (1980)
Glass/UP	Distilled water at 40 and 60°C	C	
CF/epoxy	Water and salt water at 60°C	A	Nakanishi and Shindo (1982)
Glass/polyester	Water immersion at 60°C with $\sigma < 0.3$ UTS	A	White and Phillips (1985)
Glass/polyester	Water immersion at 60°C with $\sigma = 0.6$ UTS	C	
Epoxy	97% RH, RT with $\sigma < 0.45$ UTS	LF	Henson and Weitsman (1986)
AS4/3502 gr/ep		A	
AF126-2 adhesive bond line	75–90% RH, 10–40°C	A	Althof (1979)
FM73 adhesive bond line		A	
gr/F155	95% RH at 49°C	A	Clark (1983)
gr/F185		LF	
gr/F 155-gr/F 185 lay-up		A	
PEEK	Water at 36°C	LF	Nicolais et al. 1991
	Water at 60°C	A	
	n-Heptane at 36°C	B	
	Methylene chloride at 32°C	A	
Several kinds of glass fiber in epoxy, polyester, and vinylester	Methylene chloride at 36°C	S	
	CH_2C_{12} at increasing activities	Between LF and S	
Epoxy (including 3 modifications)	Water immersion at 50°C	B	Olmos et al. (2006)
Several kinds of glass fibers and epoxy	Water at 90°C	D	Van den Emde and Van den Dolder (1991)
E-glass roving in polyester and vinylester		C	
Several kinds of glass fibers in epoxy, polyester, and vinylester	Sulfuric acid solution at 90°C	D	
Vinyl terminated polyether	Various levels of relative humidity and temperature	B	Barton and Pritchard (1994)

E-glass swirl mat/urethane matrix	Distilled water at 23 and 50°C	A	Gao and Weitsman (1998)
	Distilled water at 90°C	C	
Glass/epoxy	Distilled water at 20°C	A	Davies et al. 1998
	Distilled water at 40°C	A	
Glass/epoxy	Distilled water at 60°C	C followed by D	
Woven glass/vinylester	Water at 40 and 65°C	LF	Yoosefinejad and Hogg (1998)
	Water at 95°C	C	
Woven glass/vinylester	Water at 40, 65 and 93°C	LF (?)	
	Water at 93°C	D	
Carbon/DEG BA DCDA epoxy	60–100% RH	LF	Imaz et al. (1991)
	Immersion in distilled water	A	
Glass/epoxy	20°C } 7 years in distilled water	LF	Davis and Choqueuse, private communication
	40°C } 7 years in distilled water	A	
	60°C } 7 years in distilled water	D	
Glass/polypropylene	Distilled water at RT and 50°C	A	David et al. private communication
	Sea water at RT	LF	
Glass/epoxy pipes EP-A	23°C	LF	Yao and Ziegmann (2007)
	35 and 50°C in water	A	
Glass/epoxy pipes UP-A	23°C	A	
	35 and 50°C in water	D	
Glass/polyvinylester UP-B	23°C	A	
	35 and 50°C in water	D	
Untoughened bismaleimide	21 and 52°C	LF	Paplham et al. 1995
	90 and 100°C	A	
T300/934	45 – 75°C immersed in 90°C distilled water	A	Zhou and Lucas (1995)
		D	
FM79 adhesive (cast samples)	RH ≤95%; T ≤50°C	LF	Althof (1979)
	95% RH at T = 70°C	A or B	
FM79 adhesive (bond line)	RH ≤95%; T ≤40°C	LF	
	95% RH at T = 50°C	A	
	95% RH at T = 70°C	B	

(continued)

Table 4.1 (continued)

Material system	Exposure	Type of weight-gain data sketched in Fig. 4.2	Reference
Glass polyester and carbon polyester	Sea water at 30°C for up to 2 years	D	Kootsookos and Mouritz (2004)
Bisphenol resin A and its glass fiber composite	Water immersion at 20°C	LF	Apicella et al. 1983
	Water immersion at 90°C	D	
Bisphenol resin B and its glass fiber composite	Water immersion at 20°C	LF	
	Water immersion at 90°C	D	
Vinyl ester and its glass fiber composite	Water immersion at 20°C	LF	
	Water immersion at 90°C	C (resin) or D (composite)	
Isophtatic resin and its glass fiber composite	Water immersion at 20°C	LF	
	Water immersion at 90°C	D followed by C	
TGEBA/BF3.MEA and TGDDM/DDS Epoxies	Distilled water at 23, 50, and 80°C	A	Liu et al. (2008)
AF 126–2 adhesive (cast samples)	RH ≤95%; T ≤50°C	LF	Althof (1979)
	95% RH at $T = 70$°C	D	
AF 126–2 adhesive (cast samples)	95% RH at T 20 and 40°C	LF	
	95% RH at T 50 and 70°C	A	
Glass fiber/various molecular weight polyester resins	Water immersion at 30°C	D	Gautier et al. (1999)
Epoxy MAS	Sea water at RT	LF	Ramirez et al (2008)
	Sea water at 40°C	D	
Vinyl ester H922L	Sea water at RT	LF	
	Sea water at 60°C	A	
Vinyl ester D8084	Sea water at RT	LF	
	Sea water at 60°C	D or A	
Vinyl ester D411	Sea water at RT	LF	
	Sea water at 60°C	D or A	

[a] Exceptions to general trend, probably due to evaporation

Such changeovers follow a consistent trend and are associated with increased severity of ambient conditions, such as higher external temperature, fluid activity, or applied stress.

4.2 $D = D(T)$ Relationships

Data for the dependence of the diffusion coefficient D on temperature T are listed in Table 4.2

D (absorption) vs. D (intial) or D (desorption), D_a, D_i, D_d

Variation of D_a/D with number of wet/dry cycles of graphite/epoxy composites			
$T = 77°C$	RH ~ 75% – 4%	$D_a/D_i = 1.0; 1.4; 1.7$ (3 cycles)	Delasi et al. (1978)
$T = 77°C$	RH ~ 100% – 4%	$D_a/D_i = 1.0; 2.8; 2.8$ (3 cycles)	
$T = 44°C$	RH ~ 100% – 4%	$D_a/D_i = 1.0; 2.7$ (2 cycles)	
$T = 44°C$	RH ~ 100% – 17%	$D_a/D_i = 1.0; 2.4$ (2 cycles)	

The aforementioned variation of D with number of wet/dry cycles is supported by the results shown in Fig. 4.3 (Hahn 1987).

Concentration dependent diffusion $D = D(m)$

Data for $D = D(m)$ to MP8000 and CEL 9220 M presented for 85°C and RH = 85 and 60%. Data reduced to a form of $D(m) = D_o + a\ (m/m_\infty)$. Weight gain data predicted to follow curve "A" and match experimental results. (Chen et al. 2005)

Two phase diffusion

		D (mm^2/s)	γ (1/s)	β (1/s)	M_∞%	References
Narmco resin	At 24°C	1.05×10^{-7}	23.7×10^{-9}	4.01×10^{-9}		Carter and Kibler (1978)
Hexel T7G145/ F584-4 epoxy/ carbon	Distilled water at 70°C				0.929 (bound) 0.288 (free)	Suh et al. 2001
Carbon/Epoxy						
T800H/2500	80°C at 90%RH	4.8×10^{-6}	8.1×10^{-7}	8.3×10^{-7}	1.27	Todo et al. (2000)
	80°C immersed	3.6×10^{-6}	3.8×10^{-7}	3.2×10^{-7}	2.06	
T800H/3631	80°C at 90%RH	4.0×10^{-6}	5.5×10^{-7}	2.5×10^{-7}	1.17	
	80°C immersed	3.5×10^{-6}	3.5×10^{-7}	1.8×10^{-7}	1.57	

4.3 Saturation Weight-Gain Levels Vs. Relative Humidity

Data relating moisture saturation level m_∞ to relative humidity (RH) are given in Table 4.3.

Table 4.2 Data for moisture and fluid diffusion coefficient D in mm^2/s

Material	$A = D_o$ (mm^2/s)	$B = T$ (°K)	Reference
T300/1034	0.44	5,058	Shen and Springer (1976)
	2.28	5,554	Loos and Springer (1979)
AS/3501-5	0.44	4,768	Whitney and Browning (1978)
	6.5	5,722	Delasi and Whiteside (1978)
	28.8	6,445	Loos and Springer (1979)
T300/5208	0.57	4,993	Loos and Springer (1979)
	0.41	5,231	Augl and Berger (1976)
934 (Neat resin)	4.85	5,113	Shen and Springer (1976)
	16.4	5,992	Delasi and Whiteside (1978)
3501-5 (Neat resin)	16.1	5,690	Delasi and Whiteside (1978)
5208 (Neat resin)	2.8	5,116	Augl and Berger (1976)
	0.051	4,060	McKague et al. (1978)
	4.19	5,488	Delasi and Whiteside (1978)
T300/1034	$D = 2.1 \times 10^{-7}$ mm^2/s		Loos and Springer (1979)
	3.8×10^{-7} mm^2/s		Mohlin (1984)
	2.28×10^{-7} mm^2/s		Loos and Springer (1979)
XAS/914	2.38×10^{-7} mm^2/s		Collings and Copley (1983)
T300/epoxy	1.95×10^{-7} mm^2/s		Gilat and Broutman (1978)
T300/914C	2.10×10^{-7} mm^2/s		Rao et al. (1995)
Epoxy	ln D vs. $1/T$ (°K)		Lin and Chen (2005)
	Arrhenius plot requires two lines, with kink around 140°C		
E-glass fiber cloth/epoxy	ln D vs. T (°C)		Bonniau and Bunsell (1981)
S/3501-5	D at several temperatures between 23 and 82°C		Delasi and Whiteside (1978)
B/5505			
Nylon 6.6	Arrhenius plots		Ishak and Berry (1994)
Chopped fiber	Reinforced volume ratio		
Reinforced nylon 6.6	$V_f = 0.27$		
Polyester	At 30°C $D = 2.5 \times 10^{-7}$ mm^2/s		Gellert and Turley (1999)
Phenolic	At $D = 2.3 \times 10^{-7}$ mm^2/s		
Vinylester A	At $D = 4.3 \times 10^{-7}$ mm^2/s		
Vinylester B	At $D = 4.3 \times 10^{-7}$ mm^2/s		
Note: Diffusion with glass reinforcement may be affected by fibers and interface.			
T300/1034	Distilled water at 70°C; $D = 5.2 \times 10^{-7}$ mm^2/s		Mohlin (1984)
	RH = 95% at 70°C; $D = 3.8 \times 10^{-7}$ mm^2/s		
E-glass/vinylester	In deionized water at 23 and 60°C		Private communication
Data for neat resin, and with unidirectional, bi-directional, and tri-directional reinforcement			
BSL-313A modified epoxy	$D = 1.7 \times 10^{-7}$ mm^2/s; RH = 70%, T = 20°C		Althof (1979)
	$D = 4.4 \times 10^{-7}$ mm^2/s; RH = 70%, T = 40°C		
	$D = 1.4 \times 10^{-7}$ mm^2/s; RH = 95%, T = 20°C		

(continued)

Table 4.2 (continued)

Material	$A = D_o$ (mm^2/s) $B = T$ (°K)	Reference
	$D = 13.9 \times 10^{-7}$ mm²/s; RH = 95%, $T = 40$°C	
	$D = 15.3 \times 10^{-7}$ mm²/s; RH = 70%, $T = 50$°C	
FM79 modified epoxy	$D = 4.2 \times 10^{-7}$ mm²/s; RH = 70%, $T = 20$°C	
	$D = 11.1 \times 10^{-7}$ mm²/s; RH = 70%, $T = 40$°C	
	$D = 3.1 \times 10^{-7}$ mm²/s; RH = 95%, $T = 20$°C	
	$D = 33.3 \times 10^{-7}$ mm²/s; RH = 95%, $T = 40$°C	
AF126-2 epoxy nitril	$D = 2.2 \times 10^{-7}$ mm²/s; RH = 70%, $T = 20$°C	
	$D = 5.6 \times 10^{-7}$ mm²/s; RH = 70%, $T = 40$°C	
	$D = 1.7 \times 10^{-7}$ mm²/s; RH = 95%, $T = 20$°C	
	$D = 5.6 \times 10^{-7}$ mm²/s; RH = 95%, $T = 40$°C	
3501-5, NMD 2373, 5208, 3501–6, 3502, 934		Delasi and Whiteside (1978)
Epoxy at 65°C	$D \sim 1.9 \times 10^{-6}$ mm²/s	Grayson and Wolf (1985)
Epoxy	D vs. $1/T$ (°K)	
Woven cloth carbon fiber	Arrhenius plots	
Cross ply carbon fiber		
Carbon/epoxy	D vs. 1,000/T (°K) plots at 60%, 75%, 100% RH and immersion in distilled water	Imaz et al. (1991)
Epikote 828 epoxy	D vs. $1/T$ (°K) plot	Nicolais et al. (1978)
Vinylester/carbon	ln D vs. 1000/T (°K) plot	Aiello et al. (2006)
Bismaleimide resin	ln D vs. 1,000/T (°K) plot	Paplham et al. (1995)
T300/934	Table of D vs. T (°C) in distilled water	Zhou and Lucas (1995)
3kPETU/G40-800	Some values of D	Bullions et al. (2000)
BMI5292/IM7		
Avimid K3B/IM7		
3501-5/AS		
Boron/epoxy	Log $D = -2.6$–1,459/T (°K)	Carpenter (1973)
IM7/K3B	$D = 0.01$ exp (−4,750/T (°K))	Vanlandingham et al. (1997)
T300/914C	$D = 2.38 \times 10^{-7}$ mm²/s	Rao et al. (1995)
Different laminate lay-ups	Avg. ratio of top area to side area = 6.14	
	$D = 1.94 \times 10^{-7}$ mm²/s	
	Average ratio of top area to side area = 8.71 (suggesting wicking through side edges)	

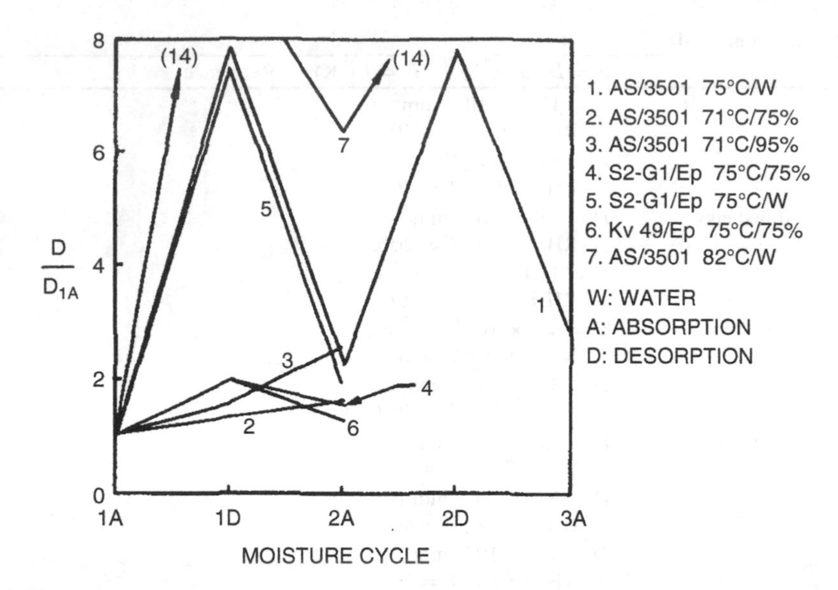

Fig. 4.3 Change of moisture diffusivity with moisture cycling (Hahn 1987)
Credit: Hahn, H. T. (1987). "Hygrothermal Damage in Graphite/Epoxy Laminates." Journal of Engineering Materials and Technology, 109(1), pp. 3–11, Fig. 4. Copyright ASME. Reproduced with permission

Table 4.3 Relations between saturation levels M_∞ and ambient relative humidity $M_\infty = \alpha(\text{RH\%})^b$ or $M_\infty = \alpha\Phi^b$ Φ = RH%: 100

Material	*a*	*b*	RH or Φ	
3501-5, 3501-6, 3502, 934, 5208	6.3	1.7	RH	Delasi and Whiteside (1978)
MD2373	9.9	2.3	RH	
AS/3501-5	1.8	1.5	RH	
B/5505	1.7	1.5	RH	
5208	7.1	1.45	Φ at 23°C $<T<$ 80°C	Wright (1981)
MY750	2.5	1	Φ at 0.2°C $<T<$ 80°C	
Bisphenol A	0.0075	1	Φ at 23°C	Menges and Gitschner (1980)
Bisphenol A/glass mat	0.004	1	Φ at 23°C	
Carbon/epoxy	0.011	1	Φ	Imaz et al. (1991)
Epoxy	0.032	1	Φ at 30°C	Nicolais et al. (1978)
T300/5208	0.0021	1.39	Φ	Long (1979)
AS/3501-5	0.0012	1.52	Φ	
5208	0.0038	1.09	Φ	
3501-5	0.0034	1.61	Φ	
3kPETU, BMI 5292/IM7, Avimid K3B/IM7, 3501-5/AS	Values of M^∞ at specific values of RH and *T*			Bullions et al. (2000)
Neat 2.5kPETU	0.0187 ± 0.0002	1	Φ	Bullions et al. (2003)

(continued)

Table 4.3 (continued)

Material	a	b	RH or Φ	
G40-800/2.5kPETU	0.0058 ± 0.0001	1	Φ	
T300 Graphite/1034 Epoxy	0.017	1	Φ	Loos and Springer (1981)
T300 Graphite/3501-5 Epoxy	0.015	1	Φ	
AS Graphite/5208 Epoxy	0.019	1	Φ	
Carbon/PMR15	0.014	1	Φ	Jacob et al. private communication
IM7/Avimid K3B	0.0011	1.34	Φ	Van Landingham (1997)
Jute/epoxy	0.00003	2.64	Φ	Rao et al. (1984)
E-glass cloth/ Bisphenol A	0.009 – 0.0101	1	Φ	Bonniau and Bunsell (1981)
PEEK	0.0075	1	Φ	Nicolais et al. (1985)
FM73 Cast adhesive $T = 20°C$	0.0023	2	Φ	Althof (1979)
FM73 Cast adhesive $T = 40°C$	0.0028	2	Φ	
FM73 Cast adhesive $T = 50°C$	0.0030	2	Φ	
FM73 Cast adhesive $T = 70°C$	0.0044	2	Φ	

Material	M_∞ data		
MY720 epoxy	3.9% at $T - 65°C$		Grayson and Wolf (1985)
Graphite fiber/ vinylester rebars	$\log M_\infty = -74.9 + 22 \times 10^3/T$ (°K)		Aiello et al. (2006)
PEEK	Exposure to CH_2Cl_2 at various activity levels $M_\infty = 0.05a + 0.5a^2$ a in decimal fraction		Nicolais et al. (1991)
Polyester	1.4%		Gellert and Turley (1999)
Phenolic	>20%	Immersion in sea water at 30°C	
Vinylester A	0.8%		
Vinylester B	0.8%		
Glass/Polyester	0.85%		
Glass/Phenolic	8.8%		
Glass/Vinylester A	0.44%		
Glass/Vinylester B	0.44%		
T300/5208	M_∞ under water immersion consistently exceeds (by 25–50%)		Long (1979)
AS/3501-5			
5208	M_∞ under exposure to RH = 95%		
3501-5			
P(AA)	Plots of M_∞ vs. RH for hydrophilic polymers		Thijs et al. (2007)
P(HEMA)			
P(VIM)			
PMMA			
P(MeOx)			
P(Etox)			
P(NIPAM)			
P(DNAEMA)			

(continued)

Table 4.3 (continued)

Material	M_∞ data		
M_∞ under cyclic exposure			
Graphite/epoxy			
$M_{\infty\alpha}$ (absorption) vs. $M_{\infty i}$ (intial) graphite/epoxy composite			
$T = 77°C$	RH ~ 75 – 4%	$M_{\infty i} = 1.03\%$, $M_{\infty\alpha} = 1.16\%$; 1.18% (3 cycles)	Delasi and whiteside (1978)[a]
$T = 77°C$	RH ~ 100 – 4%	$M_{\infty i} = 1.79\%$, $M_{\infty\alpha} = 1.79\%$; 1.8% (3 cycles)	
$T = 44°C$	RH ~ 100 – 4%	$M_{\infty i} = 1.59\%$, $M_{\infty\alpha} = 1.78\%$ (2 cycles)	
$T = 44°C$	RH ~ 100 – 17%	$M_{\infty i} = 1.59\%$, $M_{\infty\alpha} = 1.72\%$ (2 cycles)	

[a] Data plot reduced by author

4.4 Fluid Effects on the Suppression of the Glass Transition Temperature T_g

Several plots relating the glass transition temperature T_g to moisture content are shown in Figs. 4.4–4.7 below.

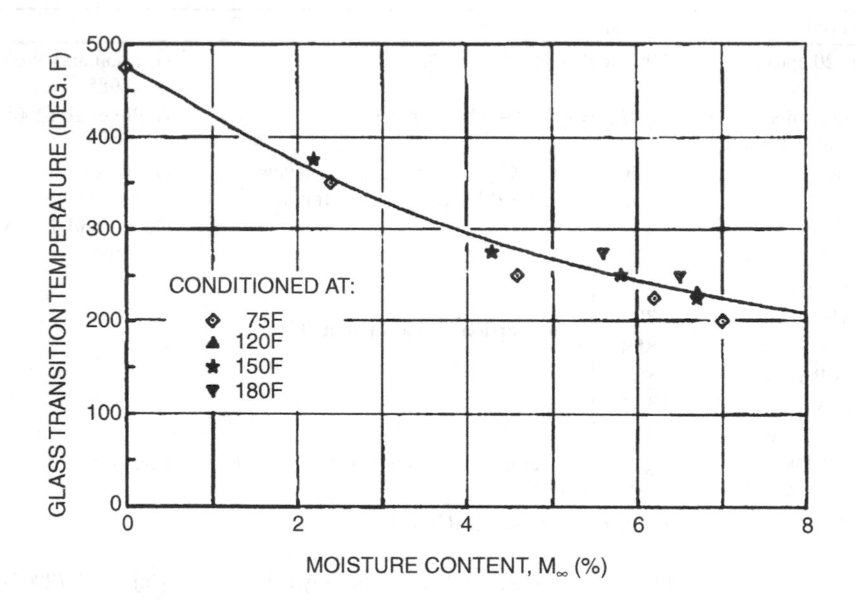

Fig. 4.4 Effect of absorbed moisture on glass transition temperature of NARMCO 5209 resin (McKague et al. 1978)

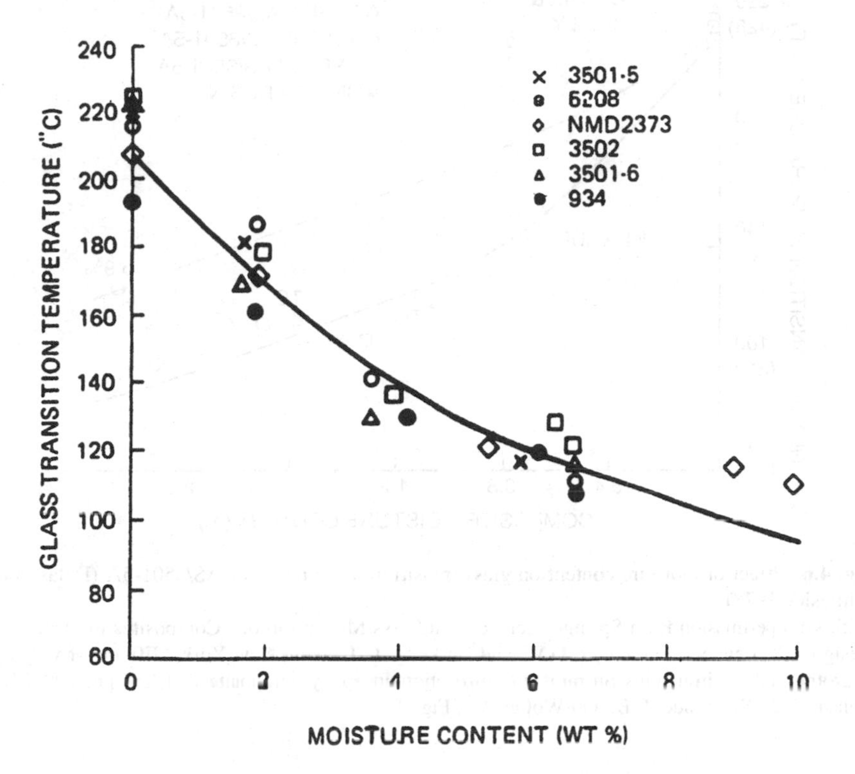

Fig. 4.5 Relationship between equilibrium moisture concentration and glass transition temperature of epoxy resins (Delasi and Whiteside 1978)
Credit Line: Reprinted, with permission, from "Effect of Moisture on Epoxy Resins and Composites." Copyright ASTM International, 100 Barr Harbor Drive, West Conshohocken, PA 19428

For various epoxy resins, the suppression of T_g in °C can be approximated by the expression (reduced from data presented by Wright 1981)

$$\Delta T_g = -Km\% \quad 13<K<28,\ K_{avg} = 19.5 \tag{4.1}$$

4.5 Fluid Induced Swelling Data

Data for fluid expansional coefficient β are listed in Table 4.4.

4.6 Records of Moisture Distribution

Moisture distribution profiles during transient states were measured by two distinct techniques. One method utilized exposure to radioactive deuterium (D_2O), in which case total weight-gain data were indiscernible from those for H_2O, and distributions

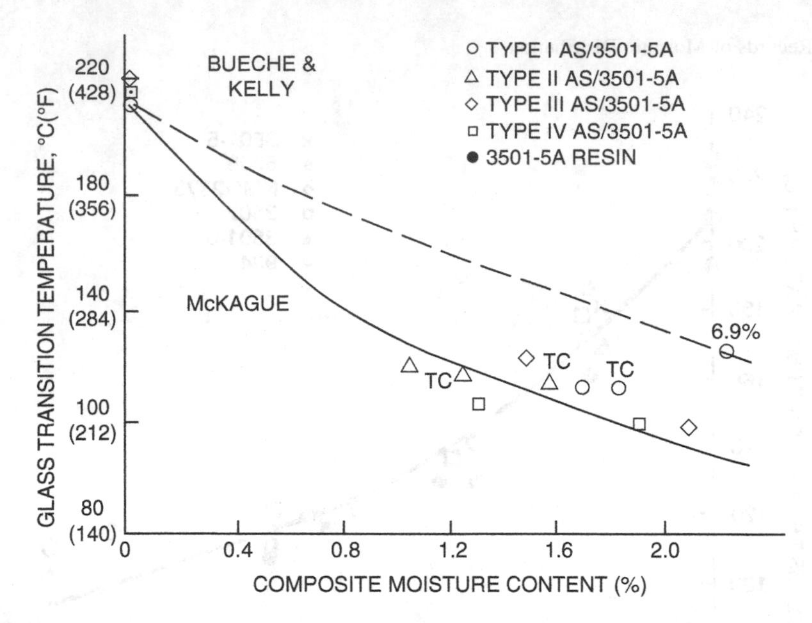

Fig. 4.6 Effect of moisture content on glass transition temperature of AS/3501-5A (Delasi and Whiteside 1978)
With kind permission from Springer Science+Business Media: Fibrous Composites in Structural Design, eds. Lenoe, E., Oplinger, D.W. and Burke, J. J., Plenum, New York, "Effects of varying hygrothermal environments on moisture absorption in epoxy composites." 1980, pp. 809–818, Delasi, R. J., Whiteside, J. B. and Wolter, W., Fig. 9

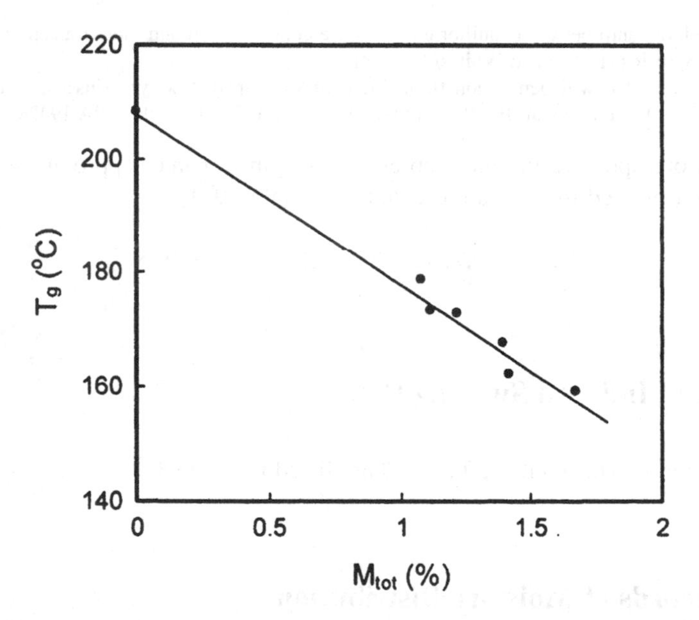

Fig. 4.7 Wet T_g depression exhibiting linear relation with equilibrium water uptake of Hexcel TG145/F584-4 carbon/epoxy composite (Suh et al. 2001)
Credit: Suh, D., Ku, M., Nam, J., Kim, B., and Yoon, S. "Equilibrium Water Uptake of Epoxy/Carbon Fiber Composites in Hygrothermal Environmental Conditions." 35(3), pp. 264–278, copyright © 2001 by *Journal of Composite Materials*. Reprinted by Permission of SAGE

Table 4.4 Swelling coefficient β (%) (per 1% moisture weight gain) β_V, β_L, and β_T are volumetric, longitudinal, and transverse coefficients, respectively

Material	β (%)		
Whenever appropriate the data are reported in the form of $\beta_v = d\,M_\infty - C$, where M_∞ in % weight gain			
3501-5	$D = 0.98$ or 0.93	$C = 0.46$ or 0.55	Delasi and Whiteside (1978)
NMD2373	0.84	0.53	
5208	0.77	0.22	
3501-6	0.87	0.61	
3502	0.85	0.61	
934	0.94	0.32	
3501-5	0.98	0.5	Wright (1981)
5208	0.75	0.5	
MY-720	0.3	0	
Bisphenol A	$\beta_v = 0.35$		Menges and Gitschner (1980)
5208	$\beta_v = 0.7$–0.75		McKague et al (1978)
Kerimid723/T300	$\beta_L = \beta_T = 0.25$		Paplham et al. (1995)
Epoxy	$\beta_L = \beta_T = 0.42$		Hahn (1987)
S2-G1/Ep	$\beta_T = 0.43$		
Kv49/Ep	$\beta_T \sim 0.43$ (approximation)		Lo et al (1982)
AS/3501	$\beta_L \sim 0$ (approximation)		
Boron/epoxy	$\beta_T = 0.168$		Chamis and Sinclair (1982)
Boron/polyimide	0.168		
T300/epoxy	0.129		
Kv49/epoxy	0.151		
AS/epoxy	0.129		
AS4/3501-6	0.22		Hooper et al. (1991a, b)
Carbon/PMR15	0.129		Jacobs et al. private communication
Phenyl ethenyl-terminated	$\beta_T = 0.0014$ (wet)		Bullions et al. (2000)
Ultem composite	$\beta_T = 0.0023$ (wet/dry)		
AS4/3501-6	$\beta_T = 0.307$		Lee and Peppas (1993)
G40-800/2.5kPETU	$\beta_T \sim (0.6\text{ m}+2\text{ m}^2) \times 10^{-3}$ m in %[a]		Bullions et al 2003
T300/934 in distilled water	$\beta_T \sim 0.75$ for $T < 75°C$ Shrinkage occurs at $T = 90°C$, and damage detected		Zhou and Lucas (1995)

[a] Data reduced by author

were recorded at various intermediate times by radioactive detection. Typical results are shown in Figs. 4.8–4.10.

Alternatively, moisture distributions were measured by employing precision abrasion mass spectroscopy (PAMS) procedures, where fine layers were abraded sequentially across the thickness of a flat carbon/epoxy specimen and analyzing the

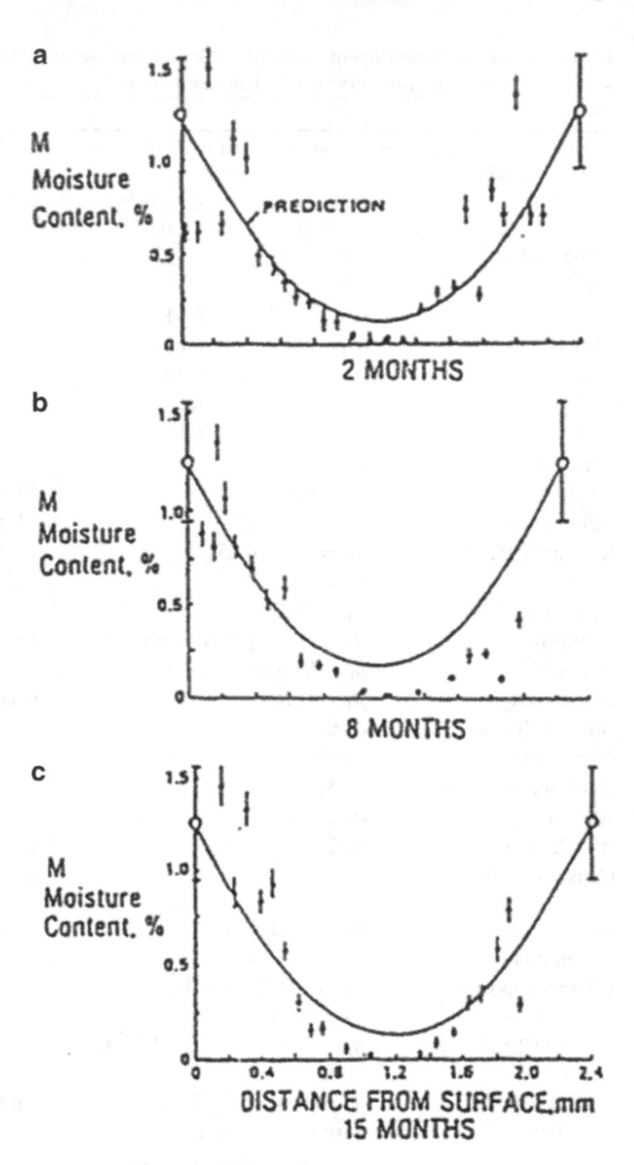

Fig. 4.8 Recorded concentration values as compared with distribution profiles predicted by Fick's law (Whiteside et al. 1984). *Dots* show average recorded concentrations of heavy water (D_2O) after 56 days of exposure to a 70%. Relative humidity of D_2O at 305 K (90°F), with *vertical lines* indicating statistical spread. Cases (**a–c**) refer to preexposure to hygrothermal cycles, after which all specimens were completely redried
Credit Line: Reprinted, with permission, from "Distribution of Absorbed Moisture in Graphite/Epoxy Laminates After Real-Time Environmental Cycling.", copyright ASTM International, 100 Barr Harbor Drive, West Conshohocken, PA 19428

released compounds. A similar, though perhaps cruder, process employed a delicate peeling of thin composite layers and recording their individual weight gains. Typical results are shown in Figs. 4.11–4.13.

Note that all the above measurements do not suffice to discriminate between Fickian and non-Fickian distribution profiles.

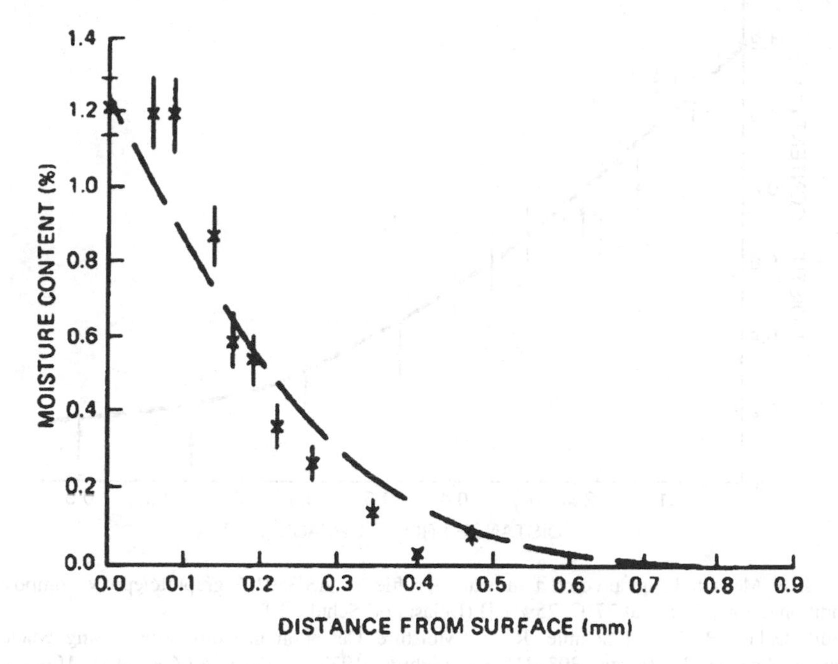

Fig. 4.9 Measured and predicted moisture profile in AS/3501-5 graphite/epoxy composite conditioned for 2.5 days at 77°C, 75% RH (Delasi and Schulte 1979)
Credit: Delasi, R. J., and Schulte, R. L. "Moisture Detection in Composites Using Nuclear Reaction Analysis." 13(4), pp. 303–310, copyright © 1979 by *Journal of Composite Materials*. Reprinted by Permission of SAGE

4.7 Moving Fluid Fronts

Distributions of H_2SO_4 and HCl across a protective polymeric layer at various times, temperatures, and concentrations are shown in Figs. 4.16 and 4.17. Note, however that the concentration fronts are somewhat slanting and not perfectly sharp (Figs. 4.14 and 4.15).

4.8 Weight Gain Data Under Cyclic Exposure

Typical absorption/desorption data for uni-directionally reinforced AS4/3502 coupons, exposed to RH = 97% to saturation and subsequently to RH = 0%, all at $T = 40°C$, are shown in Fig 4.16 below (desorption data are plotted in a reverse direction)

Note the "hysteresis loop" between the two data sets. The observation that desorption starts at a faster rate than absorption is consistent with results such as shown in Table 4.2.

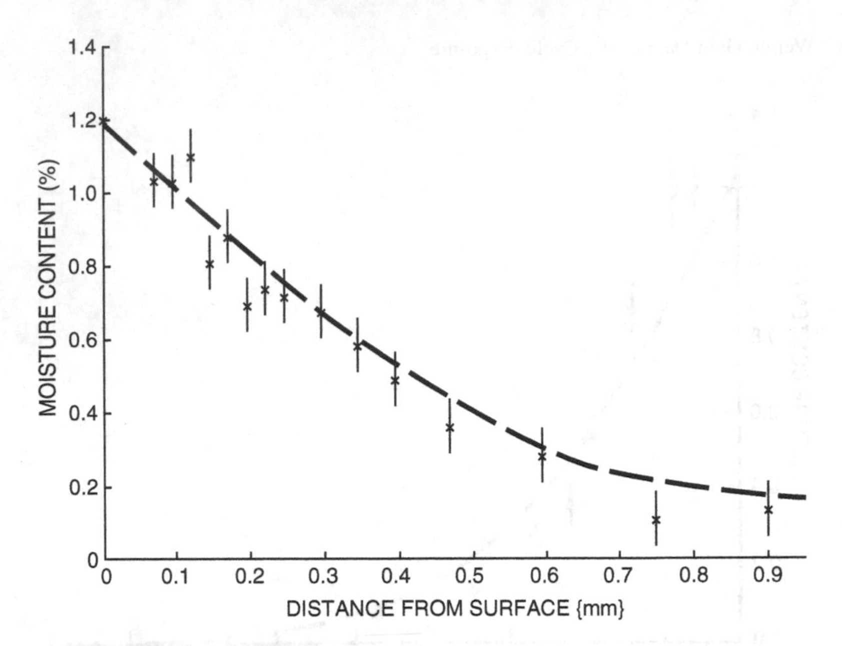

Fig. 4.10 Measured and predicted moisture profile in AS/3501-5 graphite/epoxy composite conditioned for 8.7 days at 77°C, 75% RH (Delasi and Schulte 1979)
Credit: Delasi, R. J., and Schulte, R. L. "Moisture Detection in Composites Using Nuclear Reaction Analysis." 13(4), pp. 303–310, copyright © 1979 by *Journal of Composite Materials*. Reprinted by Permission of SAGE

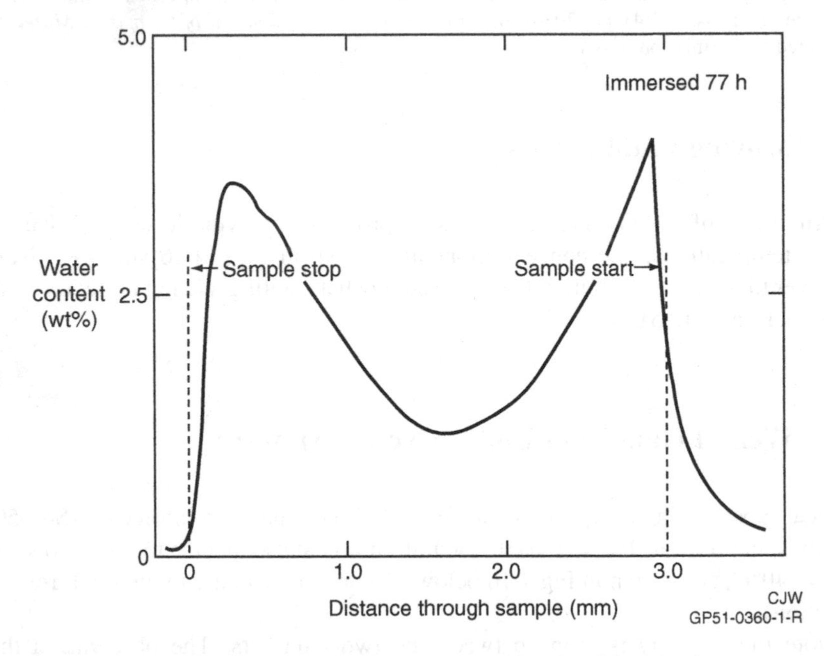

Fig. 4.11 Recorded water distribution profile as measured by a precision abration mass spectroscopy (PAMS) technique (Grayson and Wolf 1985)
Credit: Grayson, M. A. and Wolf, C. J. (1985). "Diffusion of Water in Carbon Epoxy Composite." 126, pp. 1463–1473. Copyright by TMS. Reprinted by permission of The Metallurgical Society, Inc.

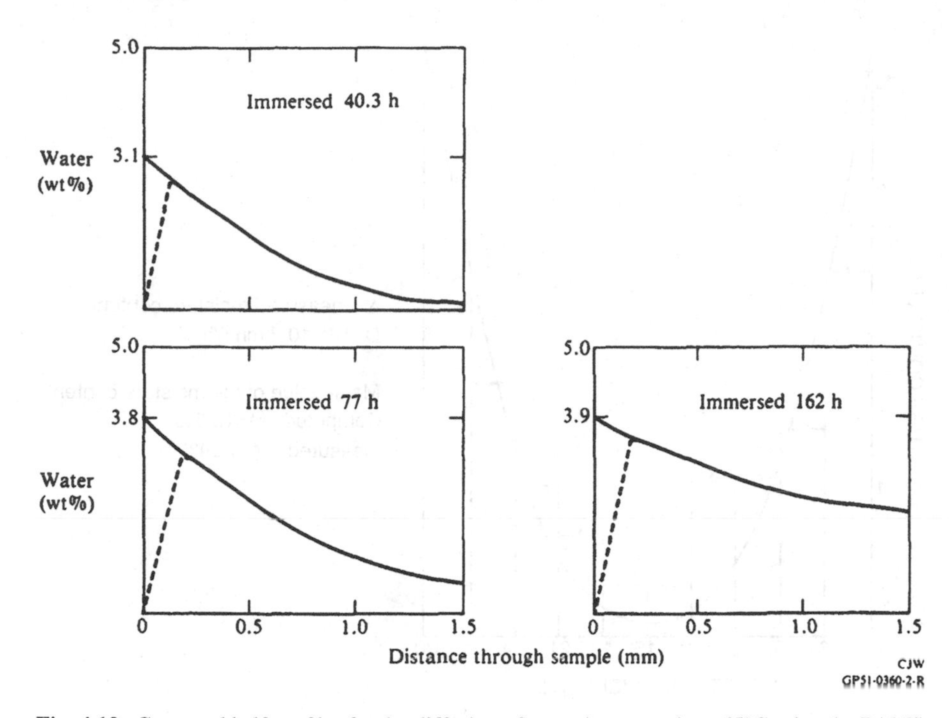

Fig. 4.12 Corrected half-profiles for the diffusion of water in neat resin at 65°C using the PAMS technique (Grayson and Wolf 1985)
Credit: Grayson, M. A. and Wolf, C. J. (1985). "Diffusion of Water in Carbon Epoxy Composite." 126, pp. 1463–1473. copyright by TMS. Reprinted by permission of The Metallurgical Society, Inc.

It is worth noting that, when compared with the data exhibited in Fig. 4.16, the absorption/desorption records for "neat" 3502 epoxy exposed in the same manner exhibit a barely discernible hysteresis as shown in Fig. 4.17 below.

Similar loops were noted for strain data recorded during absorption, desorption, and resorption of moisture under cyclic relative humidity, as shown in Fig. 4.18 (Lo et al. 1982).

Additional data were collected for anti-symmetric $[0°/90°/0_4°/90_4°/0°90°]T$ AS/3502 graphite/epoxy laminates that were exposed to three different regimes of cyclic relative humidity. As follows: (a) 9-day cycling between RH $=$ 95% and RH $=$ 0% (dry) at $T = 54°C$, (b) 16-day cycling between RH $=$ 95% and RH $=$ 0% (dry) at $T = 54°C$, and (c) 9-day cycling between RH $=$ 0% (dry) and RH $=$ 95% at 66°C. Weight gain data are shown in Fig. 4.19(a–c).

In cases (a) and (b), the samples were allowed to saturate at RH $=$ 95%, when $m_\infty = 1.025\%$ at time $t = t_s$, and subsequent weight gain data recorded at times $t - t_s$ as noted in the above figure.

Results are shown by the full dots (.), together with comparisons against the predictions of Fick's law. It can be seen that the discrepancies tend to increase with the number of cycles.

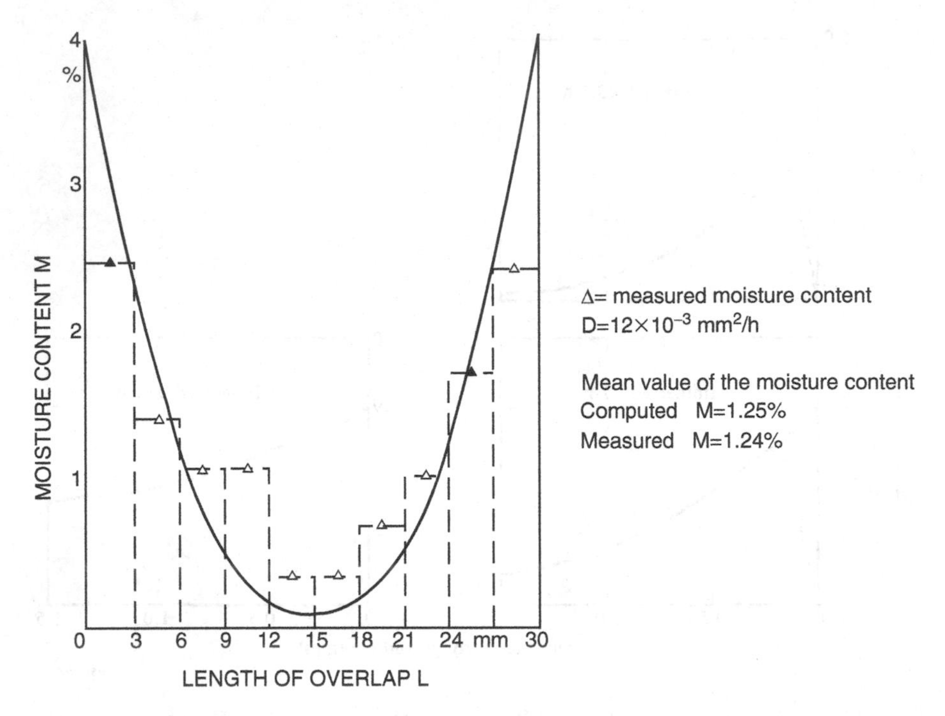

Fig. 4.13 Distribution of moisture in a FM 73 bondline after 60 days exposing to the test climate 70°C, 95% relative humidity comparison between computing and measuring (Althof 1979) Credit: Althof, W. (1979). "The Diffusion of Water Vapour in Humid Air into the Bondlines of Adhesive Bonded Metal Joints." Proceedings of the 11th National SAMPE Conference, Vol. 11, SAMPE, Azusa, November 1979, 309–332. Reprinted by permission of SAMPE

4.9 Stress Effects on Fluid Ingress and Its Relation to Damage

Weight-gain data were collected on uni-directionally reinforced AS4/3502 graphic/epoxy composite coupons loaded transversely to the fiber direction as well as on "neat" 3502 samples. All specimens were exposed to RH = 98% and $T = 40°C$, with different samples loaded in tension at 0, 15, 30, and 45% of the dry epoxy failure stress of 52 MPa.

Results for the epoxy specimens are shown in Figs. 4.20–4.24. These results show that m_∞ increases monotonically with the stress σ suggesting the relation $m_\infty \sim A\sigma^2$ that tallies with the model presented in Sect. 6.3, while D increases with σ during absorption and decreases during desorption. In addition, Fig. 4.16 demonstrates the presence of a hystersis loop in the absorption/desorption data for sufficiently high levels of stress (Fig. 4.25).

Note that the weight gain data follow curve "A" in Fig. 4.2 rather than the Fickian model. At least for the case of $\sigma = 0$ this suggests the presence of a two phase diffusion process.

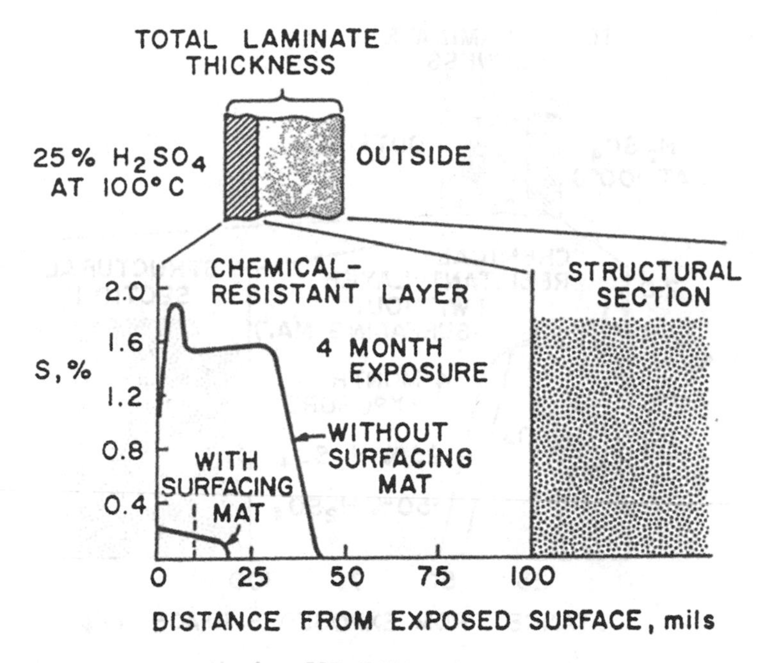

Fig. 4.14 Penetration of sulfur from 25% H_2SO_4 into glass reinforced bisphenol-A fumarate polyester laminates, with and without surfacing mat. (Regester 1969)
Credit: Regester, R. F. (1969). "Behaviour of fibre reinforced plastic materials in chemical service." *Corrosion* 25, pp. 157–167. Reprinted by permission of NACE international

Similar results obtained for the uni-directionally reinforced composite. These are shown in Figs. 4.26–4.29. Though not exhibited here, the gaps in the hysteresis loops can be shown to increase monotonically with stress levels, i.e. $\sigma = 0, 0.15, 0.30$, and 0.45 of σ_{ult}. Similarly, both m and D increase monotonically with σ during the absorption and desorption stages. The increase of m_∞ appears to be linearly related to σ, i.e. $m_\infty \sim K\sigma$.

The cyclic and stress dependent data, exhibited in Fig. 4.26–4.29 can be corroborated by observations of moisture-induced damage in fiber-reinforced composites. Several micrographs, displayed in Fig. 4.30(a–c) show the presence of multiduties of microcracks caused by exposure to ambient moisture. Although these microcracks occur at "globally" random locations, they seem to evolve about sites where adjacent fibers are positioned closely to each other and continue to meander along paths that tend to cling to the fiber/matrix interfaces. This tendency appears to be precipitated by high stress concentrations and follow weaknesses in the interphase regions. Other frequent locations for microcrack initiation and growth occur at the interfaces between fibers and resin-rich regions. This phenomenon may be attributed to high residual stresses caused by fluid induced swelling.

Though difficult to quantify, these observations strongly suggest that more damage in the form of microcracks is caused by exposure to fluctuating than constant ambient humidity, with a larger amount happening during the drying stages.

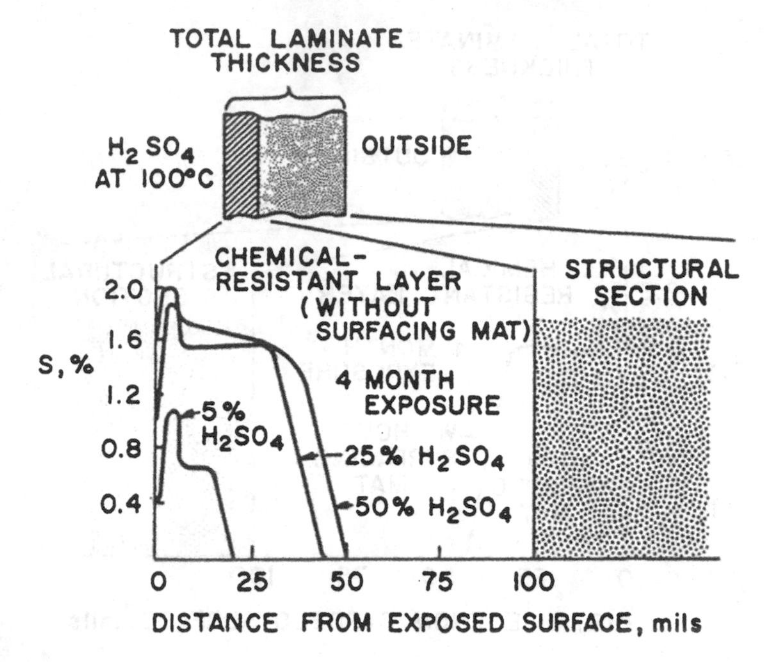

Fig. 4.15 Penetration of sulfur into glass reinforced bisphenol-A fumarate polyester laminates from sulfuric acid solutions of various concentrations (Regester 1969)
Credit: Regester, R. F. (1969). "Behaviour of fibre reinforced plastic materials in chemical service." *Corrosion* 25, pp. 157–167. Reprinted by permission of NACE international

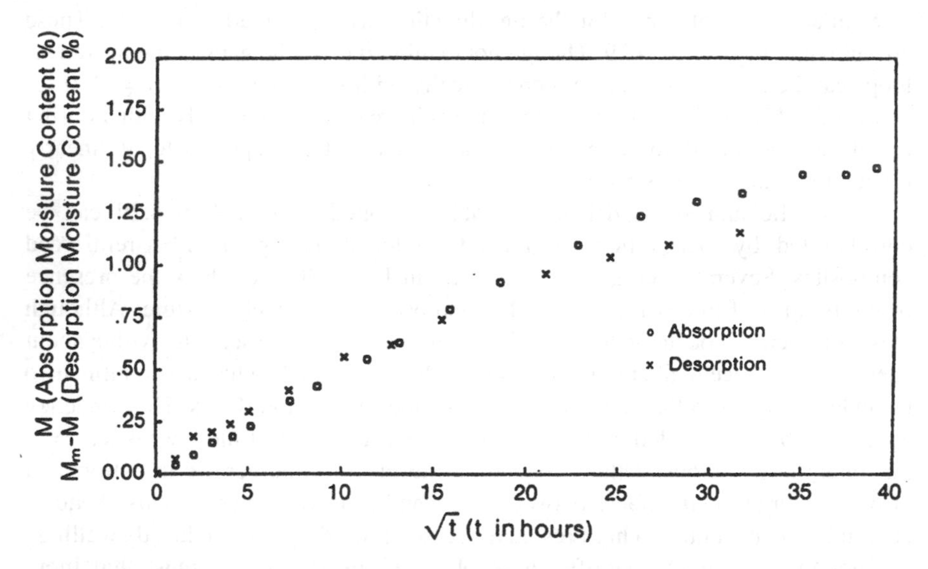

Fig. 4.16 Superimposed values of moisture content vs. $\sqrt{t}$ in unstressed reinforced AS4/3502 coupons during absorption (at RH = 97%) and desorption (at RH = 0%) at T = 40°C (Henson and Weitsman 1986)

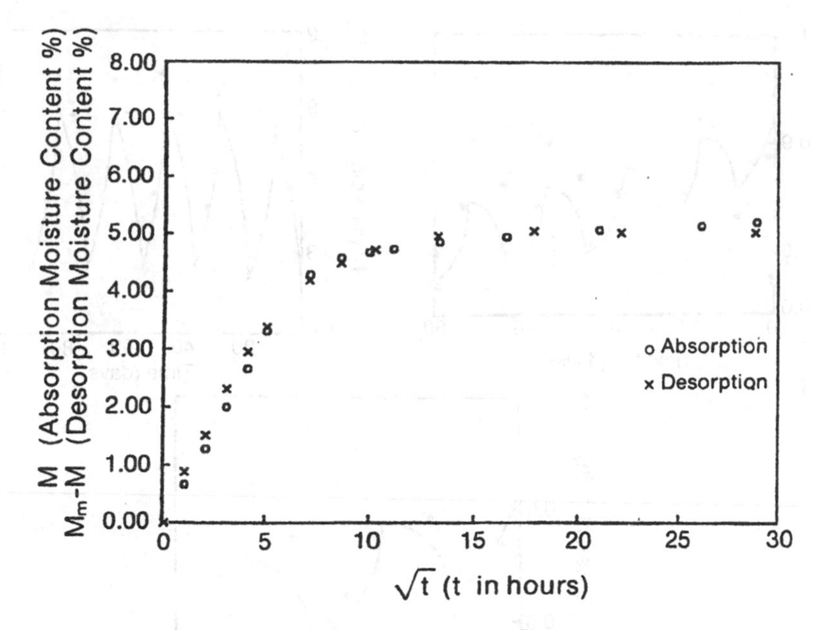

Fig. 4.17 Superimposed values of moisture content vs. $\sqrt{t}$ in unstressed 3502 epoxy coupons during absorption (at RH = 97%) and desorption (at RH = 0%) at $T = 40°C$ (Henson and Weitsman 1986)

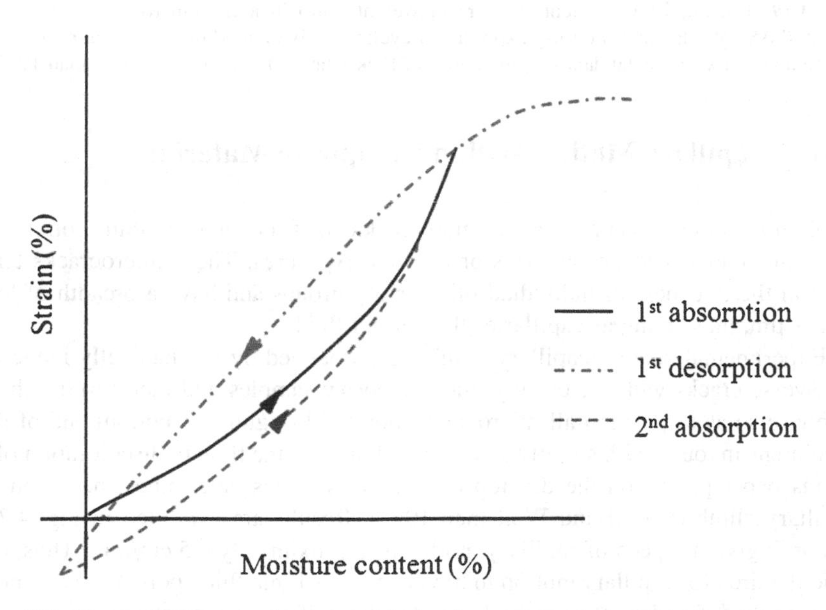

Fig. 4.18 Hysteresis loops in the moisture-induced transverse strain upon exposure to cyclic ambient relative humidity

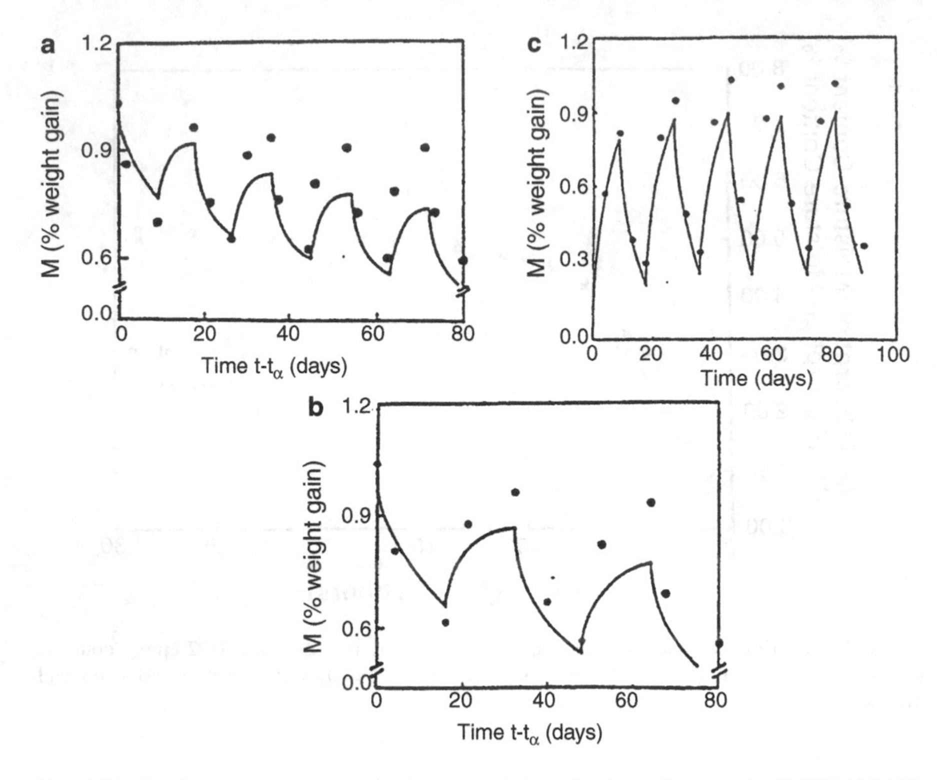

Fig. 4.19 Total moisture content (in percent weight gain) in antisymmetric [0/90/04/904/0/90] × *T* AS/3502 laminates during exposure to cyclic relative humidities. Case (**a–c**) as mentioned above. Experimental data vs. predictions of Fick's law (–). (Harper and Weitsman 1985)

4.10 Capillary Motion Within Composite Material

Capillary motion occurs inside microcracks that develop within composite laminates under mechanical loads or thermal exposures. These microcracks tend to span the thickness of individual plies or ply groups and have a breadth of less than 1 μm, thus acting as capillaries that attract fluid.

Experimental data on capillary climb were obtained by mechanically inducing transverse cracks within cross ply graphite/epoxy samples and then turning them sidewise so as to position all microcracks upward bringing the bottom end of the specimens in touch with slightly acid water. Tracking the time to discoloration of a litmus paper placed on the dry top end of the samples determines the speed of capillary climb (Kosuri and Weitsman 1995). Results are exhibited in Fig. 4.31. Accordingly, the speed of capillary motion is approximately 0.5 cm/min. Thus, the time required for capillary motion to travel across a 1 mm thick polymeric composite is about 5–6 orders of magnitude shorter than diffusion-saturation-time.

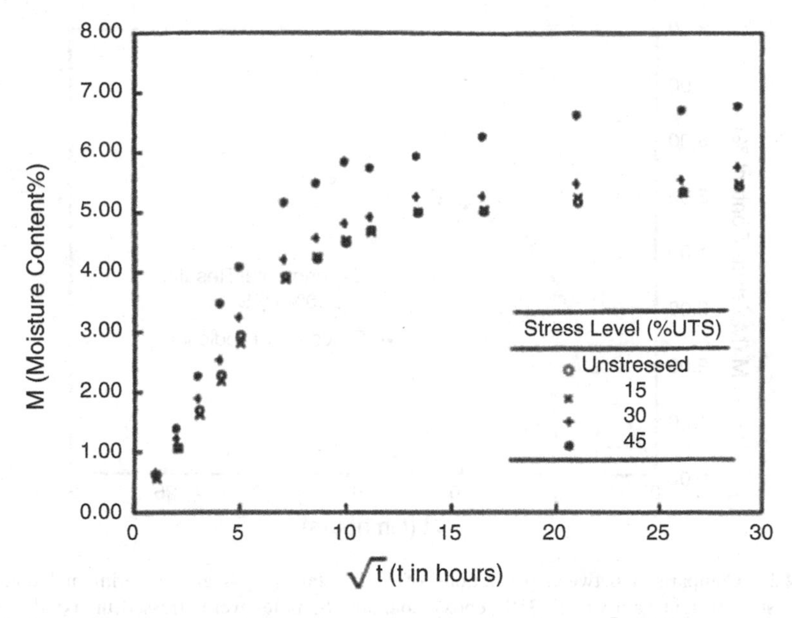

Fig. 4.20 Moisture content (average of two to four specimens) vs. $\sqrt{t}$ in 3502 epoxy coupons subjected to various stress levels during absorption: $T = 40°C$, RH 97% (Henson and Weitsman 1986)

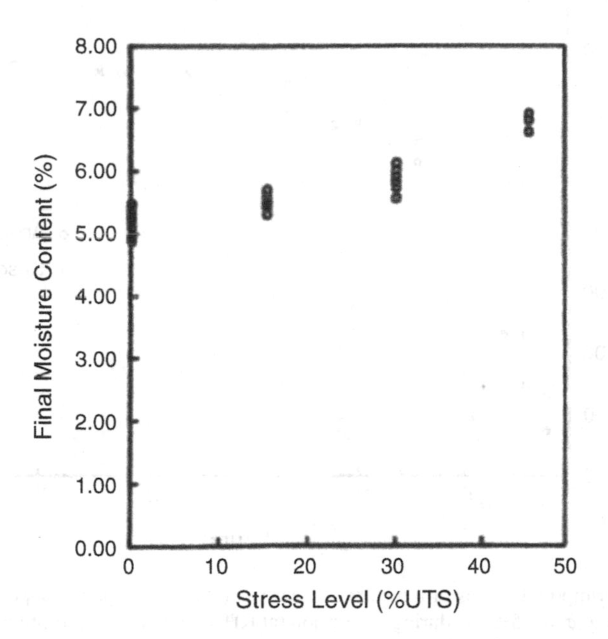

Fig. 4.21 Maximum moisture content attained in uniaxially stressed 3502 epoxy coupons, after 830 h of exposure to RH = 97% at $T = 40°C$ vs. stress level (Henson and Weitsman 1986)

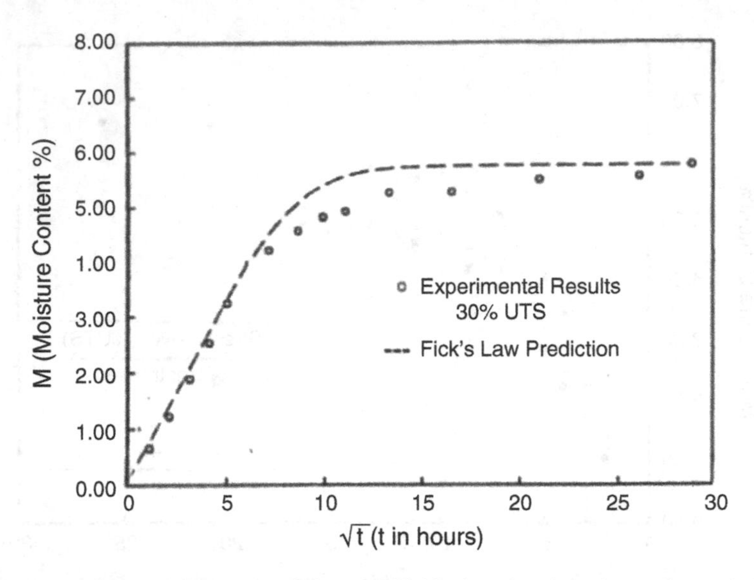

Fig. 4.22 Comparison between predictions of Fick's law (---) and experimental data (o) for moisture weight gain vs. in 3502 epoxy coupons. Samples were stressed uniaxially under $\sigma = 30\%\sigma_{ult}$. $T = 40°C$, RH 97% (Henson and Weitsman 1986)

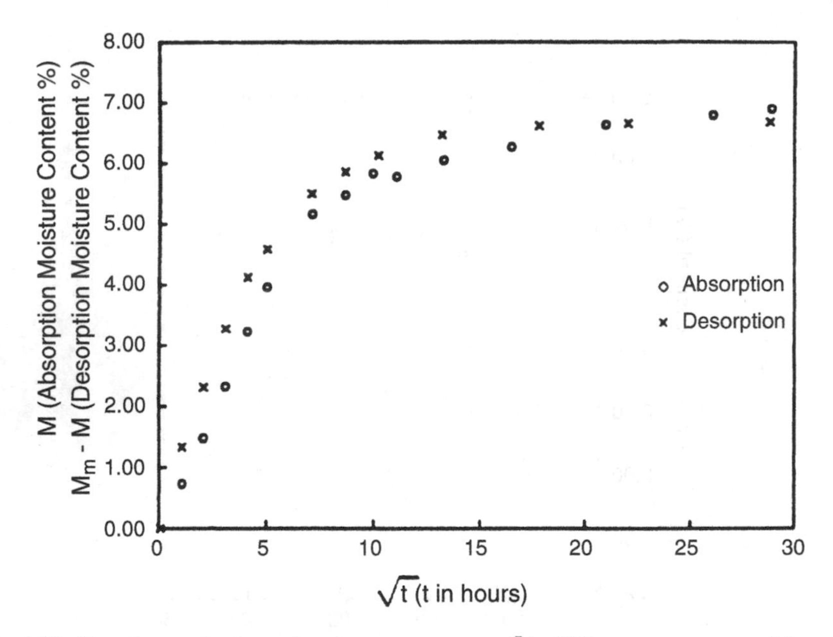

Fig. 4.23 Superimposed values of moisture content vs. $\sqrt{t}$ in 3502 epoxy coupons, subjected to uniaxial tension at $\sigma = 45\%\sigma_{ult}$, during absorption (at RH = 97%) and desorption (at RH = 0%) at $T = 40°C$ (Henson and Weitsman 1986)

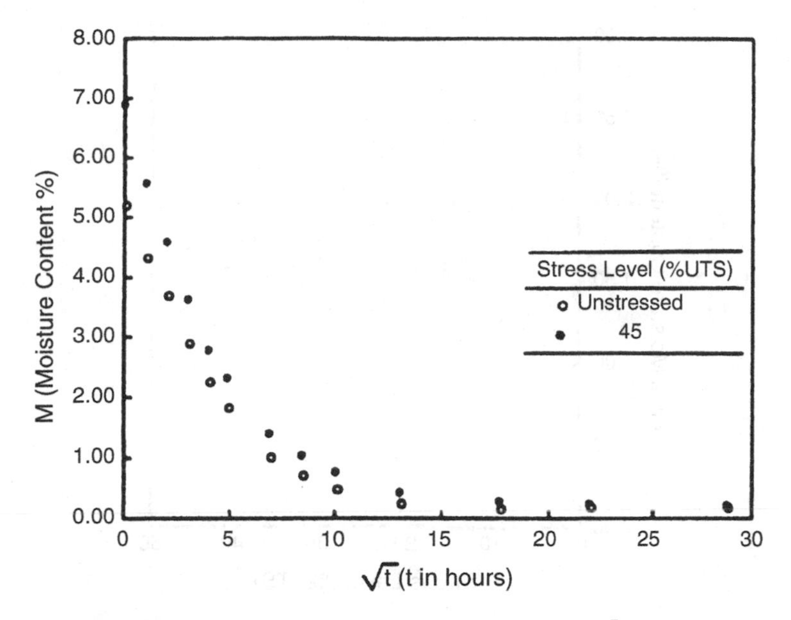

Fig. 4.24 Moisture content (average of two to four specimens) vs. $\sqrt{t}$ in 3502 epoxy coupons during desorption. Comparison between unstressed coupons and samples subjected to $\sigma = 45\%$ σ_{ult}, $T = 40°C$, RH = 0% (Henson and Weitsman 1986)

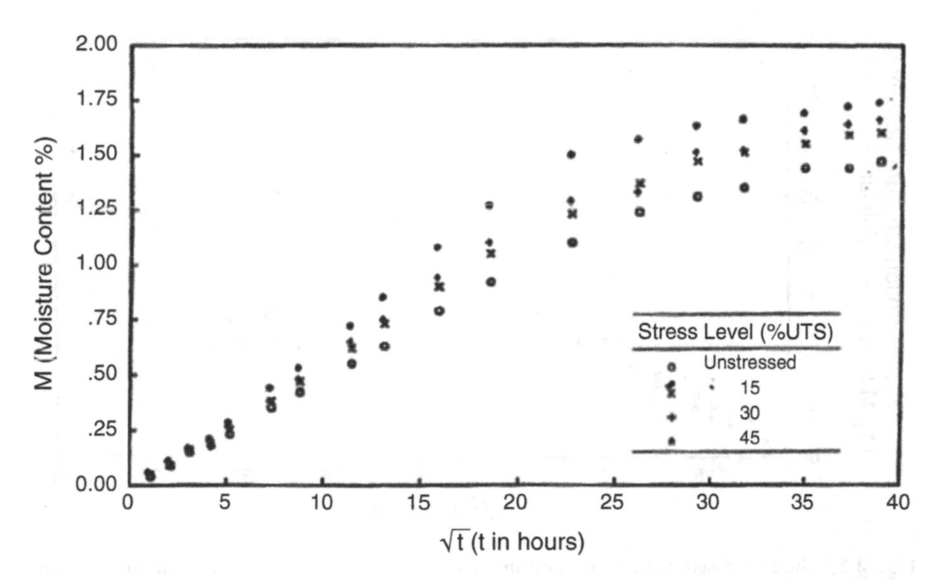

Fig. 4.25 Moisture content (average of two to four specimens) vs. $\sqrt{t}$ in unidirectionally reinforced AS4/3502 composite coupons during absorption. Coupons subjected to various levels of stress transversely to fiber direction: $T = 40°C$, RH = 97% (Henson and Weitsman 1986)

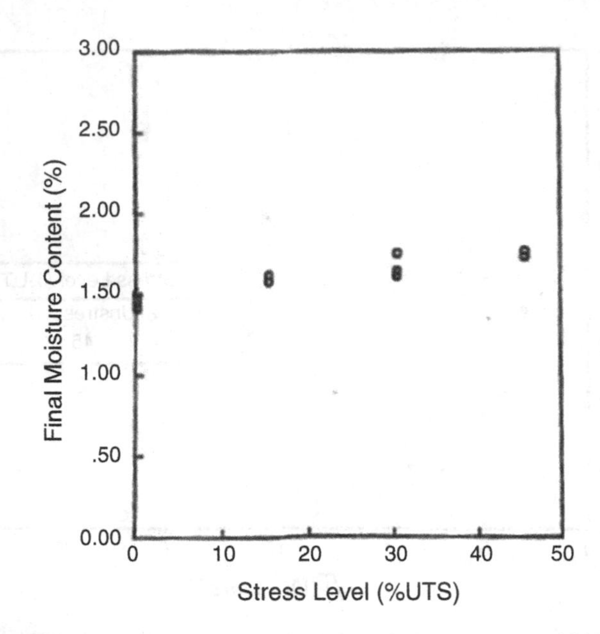

Fig. 4.26 Maximum moisture content attained in uniaxially stressed AS4/3502 composite coupons, after 1,620 h of exposure to RH = 97% at $T = 40°C$ vs. stress level (Henson and Weitsman 1986)

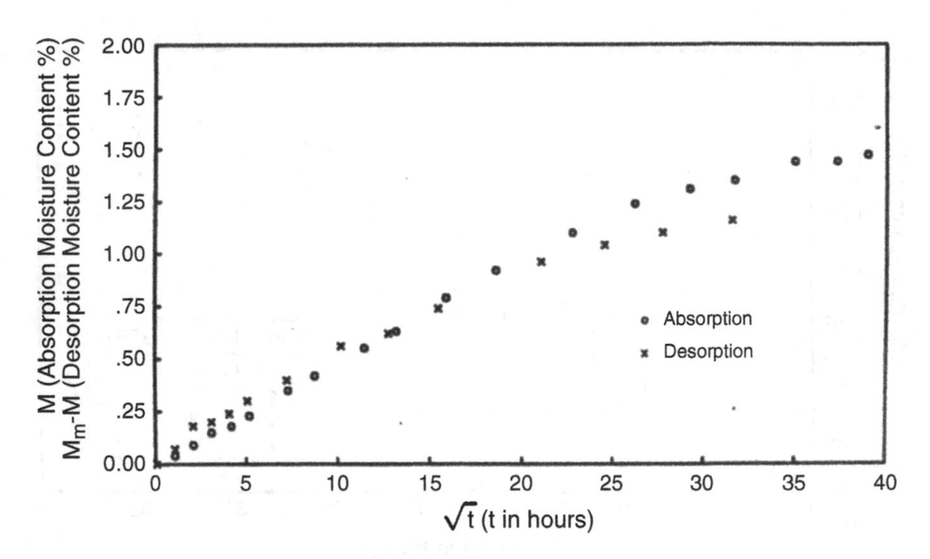

Fig. 4.27 Superimposed values of moisture content vs. $\sqrt{t}$ in unstressed, unidirectionally reinforced AS4/3502 coupons during absorption (at RH = 97%) and desorption (at RH = 0%) at $T = 40°C$ (Henson and Weitsman 1986)

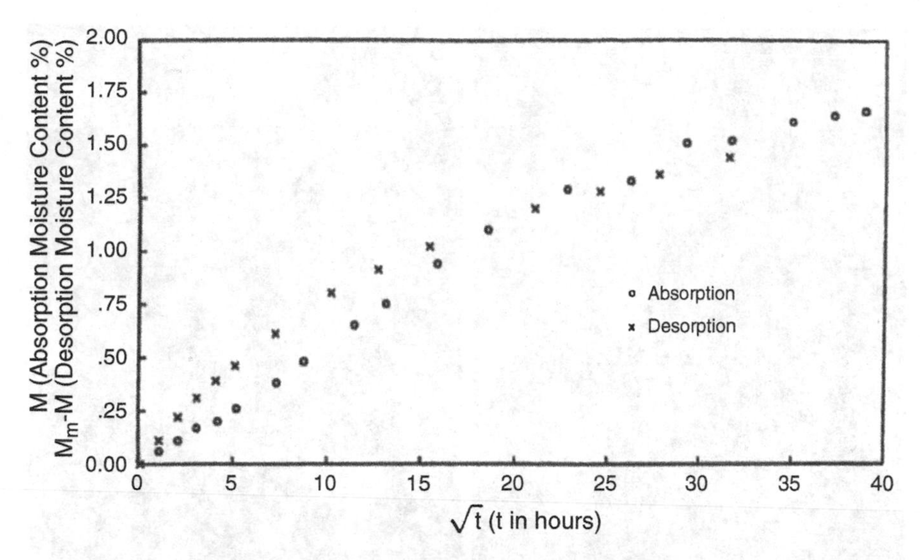

Fig. 4.28 Superimposed values of moisture content vs. $\sqrt{t}$ in unidirectionally reinforced AS4/3502 coupons during absorption (at RH = 97%) and desorption (at RH = 0%) at $T = 40°C$. Coupons subjected to uniaxial tension of $\sigma = 30\%\ \sigma_{ult}$ (Henson and Weitsman 1986)

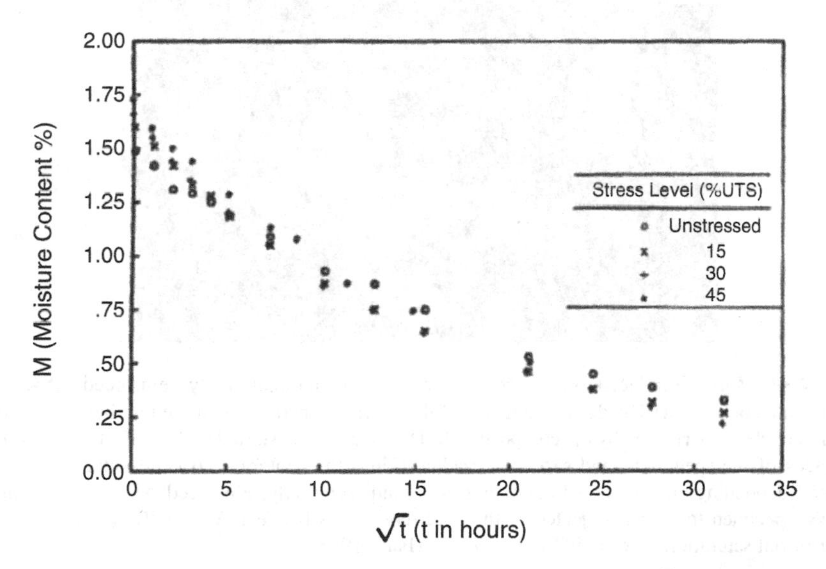

Fig. 4.29 Moisture content (average of two to four specimens) vs. $\sqrt{t}$ in unidirectionally reinforced composite coupons during desorption. Coupons subjected to various levels of stress transversely to fiber direction: $T = 40°C$, RH = 0%. (Henson and Weitsman 1986)

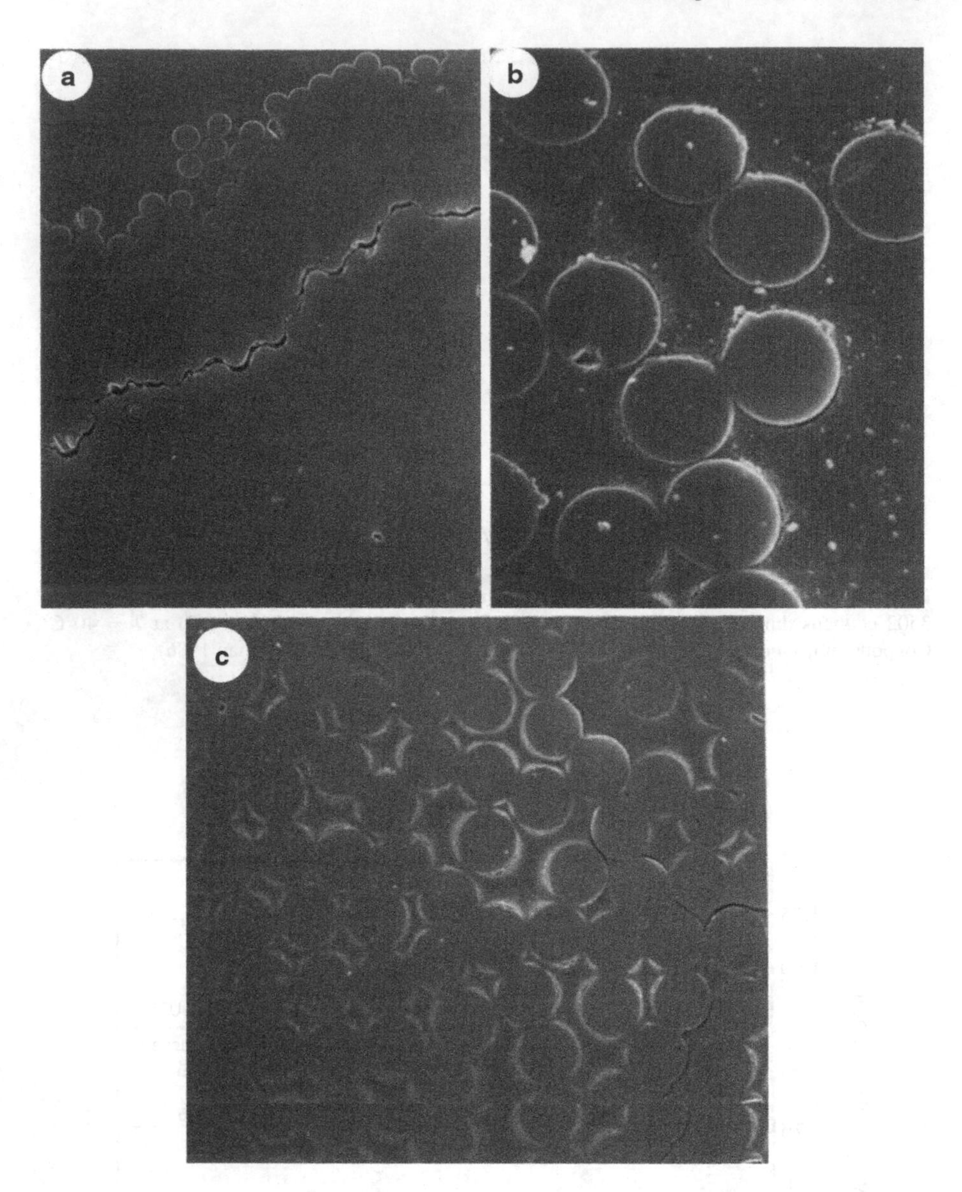

Fig. 4.30 Moisture-induced transverse microcracks in unidirectionally reinforced AS4/3502 composite coupons (**a**) The density and size of those cracks increase with the number of environmental cycles experienced by the composite. (**b**) Damage always started in the vicinity of the outer surfaces of the specimens and extended gradually into their interiors. (**c**) Individual, crescent-shaped debonding and coalesced debonding in a unidirectionally reinforced AS4/3502 graphite/epoxy specimen that was subjected to three 12-day cycles between $R_h = 65\%$ and $R_h = 95\%$, after initial saturation at $R_h = 95\%$ ($T = 73°C$) (Fang 1986)

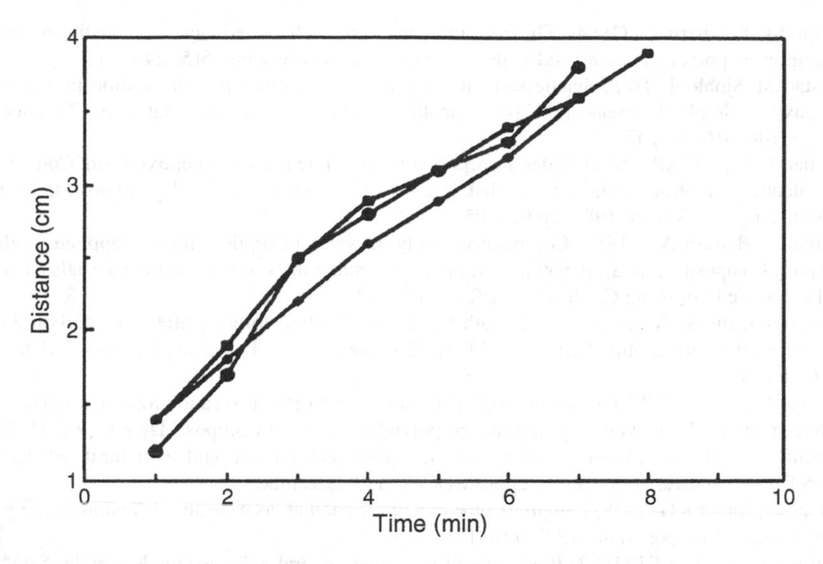

Fig. 4.31 Rate of capillary climb of sea water in precracked $[0°/90°_3]_s$ AS4/3501-6 Gr/ep coupon, recorded along the transverse cracks within the 90°~ ply group (Kosuri and Weitsman 1995)

4.11 Thermal Spiking Effects

Thermal spiking induces microcracking in composites; it is enhanced by the presence of moisture. In turn, this damage increases the moisture absorption capacity of composite lay-ups, typically by 100%. This phenomenon is of some concern in aerospace applications, where rapid temperature variations may occur during flight.

A review of earlier works on this matter was given by Hahn (1987). Some specific data are given by several authors (e.g. Collings and Stone 1985).

References

Aiello MA, Leone M, Aniskevich AN, Starkova OA (2006) Moisture effects on elastic and viscoelastic properties of CFRP rebars and vinylester binder. J Mater Civ Eng 18(5):686–691

Althof W (1979) The diffusion of water vapour in humid air into the bondlines of adhesive bonded metal joints. Proceedings of the 11th National SAMPE Conference, vol 11, SAMPE, Azusa, November 1979, pp 309–332

Apicella A, Migliaresi C, Nicolais L, Iaccarino L, Roccotelli S (1983) The water ageing of unsaturated polyester-based composites: influence of resin chemical structure. Composites 14(4):387–392

Augl JM, Berger AE (1976) The effect of moisture on carbon fiber reinforced epoxy composites. I. Diffusion. NSWC/WOL/TR 76–7. Naval Surface Weapons Center, White Oak, Silver Spring, MD

Barton SJ, Pritchard G (1994) The moisture absorption characteristics of crosslinked vinyl terminated polyethers compared with epoxies. Polym Adv Technol 5(5):245–252

Blikstad M, Sjoblonl OW, Johannesson TR (1988) Long-term moisture absorption in graphite/ epoxy angle-ply laminatesin. Environmental effects of composite materials. Technomic Lancaster, PA, pp 107–121

Bonniau P, Bunsell AR (1981) Water absorption by glass fibre reinforced epoxy resin. Composite structures. In: Proceedings of the first international conference, Paisley, Scotland; United Kingdom; 16–18 Sept, 1981, pp 92–105

Bonniau P, Bunsell AR (1984) Comparative study of water absorption theories applied to glass epoxy Composites. In: Springer GS (ed) Environmental effects on composite materials, Vol. 2, Technomic Publishing Co. Inc, CT, USA, pp 209–229

Bullions TA, Jungk JM, Loos AC, Mcgrath JE (2000) Moisture sorption effects on carbon fiber-reinforced pheny L ethnyl-terminated UltemTM composites. J Thermopl Compos Mater 13 (6):460–480

Bullions TA, Loos AC, McGrath JE (2003) Moisture sorption effects on and properties of a carbon fiber-reinforced phenylethynyl-terminated poly(etherimide). J Compos Mater 37(9):791–809

Carpenter JF (1973) Moisture sensitivity of epoxy composites and structural adhesives. McDonnell Aircraft Company, Report MDC A2640, December

Carter HG, Kibler KG (1978) Langmuir-type model for anomalous moisture diffusion in composite resins. J Compos Mater 12(2):118–131

Chamis CC, Sinclair JH (1982) Prediction of composite hygral behavior made simple. SAMPE Quarterly (ISSN 0036–0821), 14(1), 30–39

Chateauminois A, Chabert B, Soulier JP, Vincent L (1991) Hygrothermal ageing effects on viscoelastic and fatigue behaviour of glass/epoxy composites. In: Tsai SW, Springer GS (eds) Proceedings of the eighth international conference on composite materials (lCCM/8), Honolulu. Sect. 12–21: 16-E-16-E

Chen X, Zhao S, Zhai L (2005) Moisture absorption and diffusion characterization of molding compound. J Electron Packag 127(4):460–465

Clark DL (1983) Moisture absorption in hybrid composites. Texas A&M University Report MM 4665-83-16, December

Collings T, Copley S (1983) On the accelerated ageing of CFRP. Composites 14(3):180–188

Collings T, Stone D (1985) Hygrothermal effects in CFC laminates: damaging effects of temperature, moisture and thermal spiking. Compos Struct 3(3–4):341–378

Davies P, Choqueuse D, Mazeas F (1998) Composites underwater. In: Reifsnider KL, Dillard DA, Cardon AH (eds) Progress in durability analysis of composite systems. Balkema, Rotterdam, pp 19–24

Delasi R, Whiteside JB (1978) Effect of moisture on epoxy resins and composites. In: Vinson JR (ed) Advanced composite materials-environmental effects, ASTM STP 658. American Society for Testing and Materials, West Conshohocken, PA, pp 2–20

Delasi RJ, Schulte RL (1979) Moisture detection in composites using nuclear reaction analysis. J Compos Mater 13(4):303–310

Delasi RJ, Whiteside JB, Wolter W (1980) Effects of varying hygrothermal environments on moisture absorption in epoxy composites. In: Lenoe E, Oplinger DW, Burke JJ (eds) Fibrous composites in structural design. Plenum, New York, pp 809–818

Dewimille B, Thoris J, Mailfert R, Bunsell AR (1980) Hydrothermal aging of an unidirectional glass-fibre epoxy composite during water immersion. In: Bunsell AR (ed) Advances in composites materials; Proceedings of the Third International Conference on Composite Materials, Paris, France. Pergamon, Paris, pp 597–612

Fang G-P (1986) moisture and temperature effects in composite materials. Texas A&M University Report MM-5022-86-21, November 1986

French MA, Pritchard G (1991) Strength retention of glass/carbon hybrid laminates in aqueous media (chap. 46). In: Cardon AH, Verchery G (eds) Durability of polymer based composite systems for structural applications. Elsevier, Amsterdam, pp 345–354

Fukuda H (1986) Effect of moisture absorption on the mechanical proeperties of advanced composites. In: Loo TT, Sun CT (eds) Proceedings of the international symposium on composite materials and structures, Beijing., pp 50–55

Gao J, Weitsman YJ (1998) Composites in sea water: sorption, strength and fatigue. University of Tennessee Report MAES 98–4.0 CM, August 1998

Gautier L, Mortaigne B, Bellenger V (1999) Interface damage study of hydrothermally aged glass-fibre-reinforced polyester composites. Compos Sci Technol 59(16):2329–2337

Gellert EP, Turley DM (1999) Seawater immersion ageing of glass-fibre reinforced polymer laminates for marine applications. Compos A: Appl Sci Manuf 30(11):1259–1265

Gilat O, Broutman SS (1978) Effect of an external stress on moisture diffusion and degradation in graphite-reinforced epoxy laminates. In: Vinson JR (ed) Advanced composite materials-environmental effects, ASTM STP 658. American Society for Testing and Materials, West Conshohocken, PA, pp 61–83

Grayson, M. A. and Wolf, C. J. (1985). "Diffusion of water in carbon epoxy composite." *Proceedings of the Fifth International Conference on Composite Materials (lCCM/5)*, San Diego. CA. eds. Harrigan Jr., W.C., Strife J. and Dhingra A.K., Metallurgical Society of AIME, Warrendale, PA, 126, 1463–1473

Gupta VB, Drzal LT, Rich MJ (1985) The physical basis of moisture transport in a cured epoxy resin system. J Appl Polym Sci 23(4–6):4467–4493

Hahn HT (1987) Hygrothermal damage in graphite/epoxy laminates. J Eng Mater Technol 109 (1):3–11

Hamada H, Maekawa Z, Morii T, Gotoh A, Tanimoto T (1991) Durability of adhesive bonded FRP joints immersed in hot water (chap. 46). In: Cardon AH, Verchery G (eds) Durability of polymer based composite systems for structural applications. Elsevier, Amsterdam, pp 418–427

Harper BD, Weitsman Y (1985) On the effects of environmental conditioning on residual stresses in composite laminates. Int J Solid Struct 21(8):907–926

Henson MC, Weitsman YJ (1986) Stress effects on moisture transport in an epoxy resin and its composite. In: Proceedings of the third Japan-US conference on composite materials, Japan Society of Composite Materials, Tokyo, June 1986, pp 775–783

Hojo H, Tsuda K, Ogasawara K, Mishima K (1982) the effects of stress and flow velocity on corrosion behavior of vinyl ester resin and glass composites. In: Hayashi T, Kawata K, Umekawa S (eds) Proceedings of the fourth international conference on composite materials (ICCM/4), Tokyo. ISBS, Beaverton, OR, pp 1017–1024

Hooper SJ, Subramanian R, Toubia RF (1991a) Effects of moisture absorption on edge delamination, part II: an experimental study of jet fuel absorption on graphite epoxy. ASTM STP 1110:107–125

Hooper SJ, Toubia RF, Subramanian R (1991b) Effects of moisture absorption on edge delamination, part I: analysis of the effects of nonuniform moisture distributions on strain energy release rates. ASTM STP 1110:89–106

Hopfenberg HB, Frisch HL (1969) Transport of organic micromolecules in amorphous polymers. J Polym Sci B Polymer Lett 7(6):405–409

Imaz JJ, Rodriguez JL, Rubio A, Mondragon I (1991) Hydrothemal environment influence on water diffusion and mechanical behaviour of carbon fibre/epoxy laminates. J Mater Sci Lett 10 (11):662–665

Ishak ZAM, Berry JP (1994) Hygrothermal aging studies of short carbon fiber reinforced nylon 6.6. J Appl Polym Sci 51(13):2145–2155

Kootsookos A, Mouritz AP (2004) Seawater durability of glass- and carbon-polymer composites. Compos Sci Technol 64(10–11):1503–1511

Kosuri R, Weitsman YJ (1995) Sorption processes and immersed fatigue response of gr/ep composites in sea water. In: Poursartip A, Street K (eds) Proceedings of the tenth international conference on composite materials (lCCM-10), vol 4, Whistler. BC, Canada, pp 177–184

Lagrange A, Melennec C, Jacquemet R (1991) Influence of various stress conditions on the moisture diffusion of composites in distilled water and natural sea water. In: Cardon AH, Verchery G (eds) Durability of polymer based composites systems for structural applications, G, Elsevier, pp 385–392

Lee MC, Peppas NA (1993) Water transport in graphite/epoxy composites. J Appl Polym Sci 47 (8):1349–1359

Lin Y, Chen X (2005) Investigation of moisture diffusion in epoxy system: experiments and molecular dynamics simulations. Chem Phys Lett 412(4–6):322–326

Liu W, Hoa S, Pugh M (2008) Water uptake of epoxy-clay nanocomposites: experiments and model validation. Compos Sci Technol 68(9):2066–2072

Lo SY, Hahn HT, Chiao TT (1982) Swelling of KEVLAR 49/Epoxy and S2-glass/epoxy composites. In: Hayashi T, Kawata K, Umekawa S (eds) Proceedings of the fourth international conference on composite materials (ICCM/4). Progress in science and engineering of composites, vol 2, Tokyo, October 25–28, 987–1000

Long ER (1979) Moisture diffusion parameter characteristics for epoxy composites and neat resins. NASA technical paper 1474

Loos AC, Springer GS (1981) Moisture absorption of graphite-epoxy composition immersed in liquids and in humid air. In: Springer GS (ed) Environmental effects of composite materials. Technomic, Westport, CT, pp 34–49

Loos AC, Springer GS (1979) Moisture absorption of graphite-epoxy composites immersed in liquids and in humid air. J Compos Mater 13(2):131–147

Lou A, Murtha T (1988) Environmental effects on glass fiber reinforced PPS stampable composites. J Mater Eng 10(2):109–116

Ma C-CM, Huang, YH, Chang, MJ (1991) Effect of jet fuel on the mechanical properties of PPSIC. F. and Peek IC. F. After impact loading. In: Tsai SW, Springer GS (eds) Proceedings of the eighth international conference on composite materials (lCCM/8). Composites design, manufacture, and application, Honolulu, HI, pp 16-M-1–16-M-10

Manocha LM, Bahl OP, Jain RK (1982) Performance of carbon fibre reinforced epoxy composites under different environments. In: Hayashi T, Kawata K, Umekawa S (eds) Proceedings of the fourth international conference on composite materials (ICCM/4), Progress in Science and engineering of Composites, vol 2, Tokyo, October 25–28, pp 957–964

McKague EL Jr, Reynolds JD, Halkias JE (1978) Swelling and glass transition relations for epoxy matrix material in humid environments. J Appl Polym Sci 22(6):1643–1654

Menges G, Gitschner H-W (1980) Sorption behavior of glass-fibre reinforced composites and influence of diffusion media on deformation and failure behavior. In: Bunsell AR, Bathias C, Martrenchar A, Menkes D, Verchery G (eds) Proceedings of the third international conference on composite materials (ICCM/3), Paris, pp 1, 25–48

Mohlin T (1984) Deterioration in compressive performance of agraphite/epoxy laminate as a consequence of environmental exposure and fatigue loading. In: Springer GS (ed) Environmental effects of composite materials, vol 3, Technomic. Lancaster, PA, pp 163–170

Morii T, Tanimoto T, Maekawa Z, Hamada H, Kiyosumi K (1991) Effect of surface treatment on degradation behavior of GFRP in hot water. In: Cardon AH, Verchery G (eds) Durability of polymer based composite systems for structural applications. Elsevier, Amsterdam, pp 393–402

Nakanishi Y, Shindo A (1982) Deterioration of CFRP and GFRP in salt water. In: Hayashi T, Kawata K, Unlekawa S (eds) Proceedings of the fourth international conference on composite materials (ICCM/4), vol 2, Tokyo. ISBS, Beaverton, OR, pp 1009–1016

Nicolais L, Apicella A, Del Nobile MA, Mensitieri G (1991) Solvent sorption synergy in PEEK. In: Cardon AH, Verchery G (eds) Durability of polymer based composite systems for structural applications. Elsevier Applied Science, New York, pp 99–115

Nicolais L, Drioli E, Hopfenberg H, Caricati G (1978) Diffusion-controlled penetration of polymethyl methacrylate sheets by monohydric normal alcohols. J Membr Sci 3(2):231–245

Olmos D, López-Morón R, González-Benito J (2006) The nature of the glass fibre surface and its effect in the water absorption of glass fibre/epoxy composites. The use of fluorescence to obtain information at the interface. Compos Sci Technol 66(15):2758–2768

Paplham WP, Brown RA, Salin IM, Seferis JC (1995) Absorption of water in polyimide resins and composites. J Appl Polym Sci 57(2):133–137

Ramirez F, Carlsson L, Acha B (2008) Evaluation of water degradation of vinylester and epoxy matrix composites by single fiber and composite tests. J Mater Sci 43(15):5230–5242

Rao RMVGK, Shylaja Kumari HV, Raju KS (1995) Moisture diffusion behaviour of T300-914C laminates. J Reinforc Plast Compos 14(5):513–522

Rao R, Balasubramanian N, Chanda M (1984) Factors affecting moisture absorption in polymer composites part I: influence of internal factors. J Reinforc Plast Compos 3(3):232–245

Regester RF (1969) Behaviour of fibre reinforced plastic materials in chemical service. Corrosion 25:157–167

Shen CH, Springer GS (1981) Effects of moisture and temperature on the tensile strength of composite materials. In: Springer GS (ed) Environmental effects of composite materials. Technomic, Lancaster, PA, pp 15–33

Shen CH, Springer GS (1976) Moisture absorption and desorption of composite materials. J Compos Mater 10(1):2–20

Shirrell CD (1978) Diffusion of water vapor in graphite/epoxy composites. Adv Compos Mater Environ Effects, ASTM STP 658:21–42

Springer GS, Sanders BA, Tung RW (1981) Environmental effects on glass fiber reinforced polyester and vinylester composites. In: Springer GS (ed) Environmental effects of composite materials. Technomic, Westport, CT, pp 126–144

Springer GS (1981) Environmental effects on composite materials. Technomic, Westport, CT

Suh D, Ku M, Nam J, Kim B, Yoon S (2001) Equilibrium water uptake of epoxy/carbon fiber composites in hygrothermal environmental conditions. J Compos Mater 35(3):264–278

Thijs HML, Becer CR, Guerrero-Sanchez C, Fournier D, Hoogenboom R, Schubert US (2007) Water uptake of hydrophilic polymers determined by a thermal gravimetric analyzer with a controlled humidity chamber. J Mater Chem 17(46):4864–4871

Todo M, Nakamura T, Takahashi K (2000) Effects of moisture absorption on the dynamic interlaminar fracture toughness of carbon/epoxy composites. J Compos Mater 34(8):630–648

Van den Emde CAM, Van den Dolder A (1991) Comparison of environmental stress corrosion cracking in different glass fibre reinforced thermoset composites. In: Cardon AH, Verchery G (eds) Durability of polymer based composite systems for structural applications). Elsevier Applied Science, New York, pp 408–417

Vanlandingham MR, McKnight SH, Palmese GR, Eduljee RF, Gillespie JW, McCulough JRL (1997) Relating elastic modulus to indentation response using atomic force microscopy. J Mater Sci Lett 16(2):117–119

White RJ, Phillips MG (1985) Environmental stress-rupture mechanisms in glass fiber/polyester laminates. In: Harrigan WC Jr, Strife J, Dhingra AK (eds) Proceedings of the fifth international conference on composite materials (ICCM/5), San Diego, CA. Metallurgical Society of AIME, Warrendale, PA, pp 1089–1099

Whiteside JB, Delasi RJ, Schulte RL (1984) Distribution of absorbed moisture in graphite/epoxy laminates after real-time environmental cycling. In: O'Brien TK (ed) Long term behavior of composites, ASTM STP 813. American Society for Testing and Materials, Philadelphia, pp 192–205

Whitney JM, Browning CE (1978) Some anomalies associated with moisture diffusion in epoxy matrix composite materials. In: Vinson JR (ed) Advanced composite materials-environmental effects, ASTM STP 658. American Society for Testing and Materials, Philadelphia, pp 43–60

Wright W (1981) The effect of diffusion of water into epoxy resins and their carbon-fibre reinforced composites. Composites 12(3):201–205

Yao J, Ziegmann G (2007) Water absorption behavior and its influence on properties of GRP pipe. J Compos Mater 41(8):993–1008

Yoosefinejad, A., Hogg, P. J. (1998). "Durability and damage tolerance of FRCs exposed to aggressive environment." In: Visconti CI (ed) Proceedings of ECCM eighth European conference on composite materials, vol 1, Naples, Woodhead Publishing, Italy, pp 151–156

Zhou J, Lucas JP (1995) The effects of a water environment on anomalous absorption behavior in graphite/epoxy composites. Compos Sci Technol 53(1):57–64

Chapter 5
Diffusion Models

5.1 Two Phase Diffusion: Additional Considerations

As already noted in Chap. 3 a two-phase diffusion can be modeled after the assumption that the fluid penetrant consists of bound and free (mobile) phases (Carter and Kibler 1978).

The presence of a bound phase is attributed to the hydrogen bonding between the fluid and the polymeric molecules mentioned in Sect. 3.1 and perhaps also to the clustering of fluid molecules (Rogers 1965).

In an early formulation of two liquid phases (Vieth and Sladek 1965) it was considered that the fluid consisted of a diffusing phase C_{D} related by means of Henry's law and of a hole filling portion C_{H}, expressed by a Langmuir-type sorption process, namely

$$C_{\mathrm{H}} = C_{H}^{'}bp/(1 + bp) \tag{5.1}$$

and

$$C_{\mathrm{D}} = k_{\mathrm{D}}p, \tag{5.2}$$

where p denotes vapor pressure and the other symbols represent measurable material properties (with recorded values reported for the sorption of carbon dioxide in PEPT films).

Eliminating p between the above expressions it is possible to relate C_{H} to C_{D} and, upon assuming that flux is governed by C_{D} alone, arrive at a diffusion equation that is non-linear in C_{D}.

Three numerical finite difference solutions were generated for one-dimensional diffusion in an extended plate that, upon a suitable selection of material parameters, agreed reasonably well with experimental data.

Y.J. Weitsman, *Fluid Effects in Polymers and Polymeric Composites*,
Mechanical Engineering Series, DOI 10.1007/978-1-4614-1059-1_5,

Several ramifications of the above formulation are presented in a subsequent work (Vieth et al. 1976) that contains a list of related publications up to that date. It was shown that for low vapor pressure, the diffusion equation for C_D reduces to the classical Fickian form.

In addition, relationships were established between solubility vs. pressure and relative humidity. It is worth noting that weight gain predictions appear to follow curve "A" in Fig. 4.1, corresponding to a monotonic increase in fluid uptake.

The formulation presented in Sect. 3.6 (Carter and Kibler 1978) contains the crucial additional assumption of a time-dependent exchange between C_H and C_D. As before, weight gain data followed curve "A."

Shopov et al. (1996) considered several generalizations of the latter model. These included a non-linear modification of (3.28) to read

$$\frac{\partial m_b}{\partial t} = \gamma (m_m)^a - \beta (m_b)^c \tag{5.3}$$

and several variants thereof.

Several consequences of their formulations are discussed in detail, among which are two important criteria for testing the validity of the Carter–Kibler model. These state that all predicted weight gain data should scale by t^* as well as by the amplitude of the relative humidity. In addition, they demonstrate that for some combinations of their parameters it is possible to show that while the overall form of the weight gain data agrees with the Fickian predictions, the sorption process is nevertheless non-Fickian.

5.2 Two-Stage Diffusion

Two-stage diffusion is considered to correspond to weight gain data sketched by curve "B" in Fig. 4.2. Such a process accords with fluid ingress that attains an apparent equilibrium level at a shorter duration but subsequently continues to build up until attaining a final stable magnitude.

This circumstance is commensurate with diffusion coupled with viscoelastic material response. Consequently, all models that attempt to predict this phenomenon incorporate some additional time-dependent aspect within Fick's law.

It was observed that in some circumstances exposure to ambient fluid is initiated by the formation of a thin molecular layer that adheres to the exposed boundary of the polymer and that the diffusion process involves some time delay before proceeding at full capacity (e.g., Long and Richman 1960). This observation motivated Long and Richman (1960) to employ a time-dependent boundary condition even under exposure to constant environment, namely

$$m(\pm L, t) = [m_0 + m_1(1 - e^{-\beta t})]H(t) \tag{5.4}$$

in place of $m(\pm L,t) = m_0 H(t)$.

An improved match against specific data was obtained by extending (5.4) to include a spectrum of "retardation times" (Cai and Weitsman 1994), namely

$$m(\pm L, t) = \left[m_0 + \sum_{n=1}^{N} m_n (1 - e^{-\beta_n t}) \right] H(t) \tag{5.5}$$

consequently one gets

$$M(t) = m_0 M_{\text{h}}(t) + \sum_{n=1}^{N} m_n \widehat{M}(t, \beta_n), \tag{5.6}$$

where $M_{\text{h}}(t)$ is expressed by (3.11) or (3.12) and the form $\widehat{M}(t, \beta_n)$ is well known (Crank 1980, p. 53, (4.29)) and reproduced in that book.

Upon selecting values of β_n to spread over the time span of interest, the values of m_0 and m_n ($n = 1,2,\ldots,N$) were determined by a numerical scheme that matched the data of $M(t)$ at selected times t_k($k = 1,2,\ldots,K$). The value of D was selected from the early weight gain data. Results are shown in Fig. 5.1.

Another model (Roy et al. 2000) is based upon the assumption that the time dependence resides in the diffusion coefficient D, with the generalization that, due to the time-temperature analogy, depends also on the temperature T. Thus, in analogy with the common viscoelastic formulation D is expressed by a Prony series, namely

$$D = D(t) = D_0 + \sum_{r=1}^{R} D_r (1 - e^{-t/\tau_r}), \tag{5.7}$$

with temperature dependence considered to dwell in D_0 and D_r.[1]

The weight gain M(t) is again given by (3.11) but with t replaced by the integrated form of (5.7) with respect to time.

Upon selection of a specific spectrum of retardation times $\tau_{\gamma,}$ the remaining parameters were determined by a numerical approximation technique, so as to match epoxy weight gain data at four levels of temperature. It is interesting to note that the forms of those data seem to have followed curve "B" in Fig. 4.2 for $T = 23$, 50, and 60°C but switched to curve "C" at $T = 70$°C.[2]

A different approach attributes two-stage diffusion to the combined effect of two time-dependent phenomena, namely those of diffusion time t_{D} and polymeric

[1] This form is akin to that developed by Jackle and Frisch (1986).

[2] Note that the history dependence of D was suggested by basic thermodynamic considerations, to be discussed in Chap. 6.

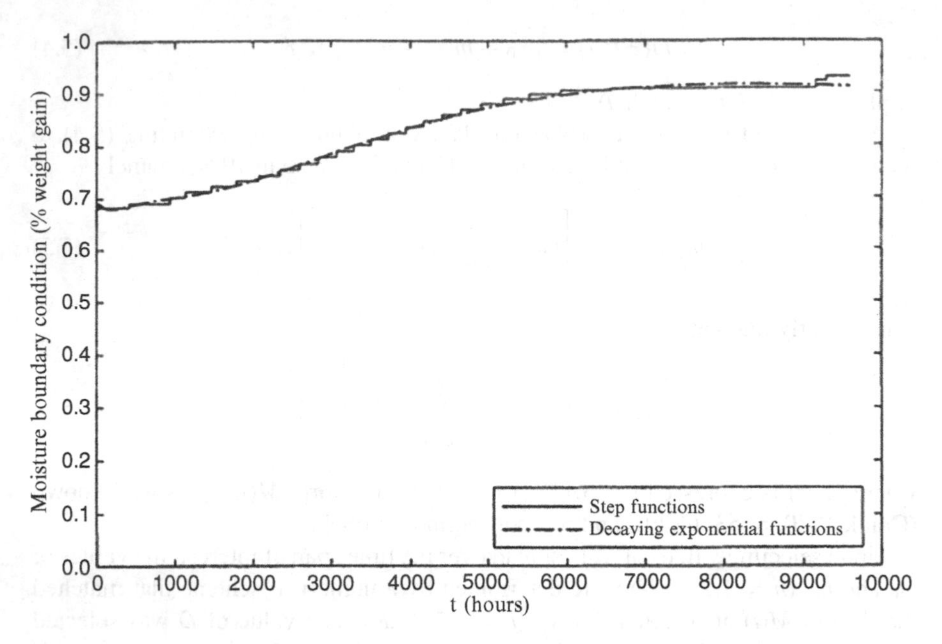

Fig. 5.1 The boundary condition represented by a series of step functions and by a series of decaying exponential functions (Cai and Weitsman 1994)

viscoelasticity. The latter may be considered to assume a single typical value of τ_v of viscoelastic time delay.

As noted earlier, the interaction between those times depends on the thickness L of the polymeric samples since t_D is proportional to L^2, while the effect of the time delay τ_v may not apply simultaneously across the entire thickness.

Intermediate values of t_D/τ_v may results in weight gain data that exhibit staggered time effects, where the role of viscoelasticity is noted only after the passage of time, i.e., for $t > \tau_v$.

Formulations suggestive of weight gain data sketched by curve "B" in Fig. 4.2 that accounts for the combined effects of diffusion and relaxation in the additive form, namely

$$M(t) = M_D(t) + M_R(t) \tag{5.8}$$

were considered by Berens (1977), Berens and Hopfenberg (1978). Note that the above form discards interactions between the above phenomena.

$M_D(t)$ was expressed by (3.11), while $M_R(t)$ was assumed to be given by

$$M_R(t) = \sum_{i=1}^{N} m_i^{Re}(1 - e^{-w_i t}), \tag{5.9}$$

where m^{Re} denotes equilibrium value.

A modified version of (5.8) can be employed when fitting it against experimental data (Bond 2005), namely

$$M(t) = \Phi M_D(t) + (1 - \Phi)M_R(t). \tag{5.10}$$

A suitable selection of Φ should match two-stage diffusion data. If the second stage is barely noticeable until time $t = t_v$ it may well be that replacing $M_R(t)$ by $M_R(t - t_v)H(t - t_v)$ in (5.10) would provide a yet better agreement.

It was observed (Bond 2005) that the two-stage model, with just one retardation time considered in (5.9), approximates the two-phase model for large values of t^*.

A comparison between yet another form of weight gain data that may resemble curve "S" in Fig. 4.2 was shown (De Wilde and Frolkovic 1994) to concur with a surface evaporation boundary condition (Crank 1980)

$$-D\frac{\partial m}{\partial x} = \alpha(m_0 - m_s). \tag{5.11}$$

In the above, m_s is the actual boundary concentration within the material, while m_0 corresponds to the concentration required to maintain equilibrium within the ambient region.

By properly adjusting the values of α and m_s, the authors obtained an excellent match with data that seem to follow an "S" shape weight-gain values, as shown in Fig. 5.2 below.

Alternately, it turned out that a combination of boundary condition (5.4) to generate $M_D(t)$ with a relaxation component $M_R(t)$ that contains a single retardation time (De Wilde and Shopov 1994) also resulted in $M(t)$ that followed an "S" shape form. As before, it was possible to generate a nearly perfect match with data.

Given the above option for modeling weight gain data that follows the form "S," the choice should follow an examination if condition (5.11) applies to the circumstance at hand.

5.3 Free Volume Dependence

As noted in Sect. 3.1, the free volume is one of the paramount parameters that affect the ingress of fluids in polymers.

A review of models for the diffusion coefficient D (Lefebvre 1987) presented several forms that related D to the free volume V_f. These were attributed to Turnbull and Cohen (1961), Macedo and Litovitz (1965), and Doolittle and Doolittle (1957) associated with the WLF equation (Williams et al. 1955).

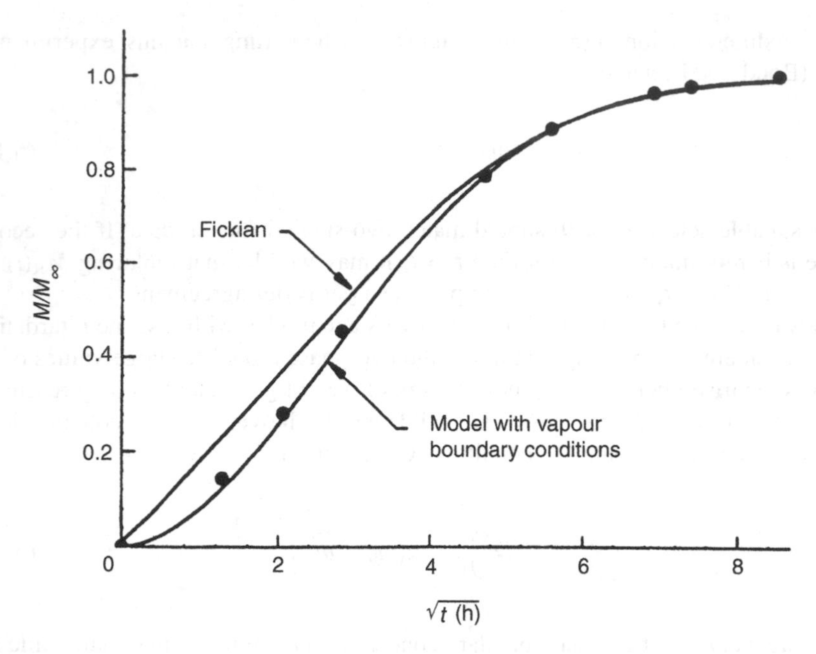

Fig. 5.2 Comparison of experimental results (*filled circle*) with proposed model (5.11) and Fickian curve. Reprinted from De Wilde and Frolkovic (1994), Copyright (1994), with permission from Elsevier

5.4 Concentration Dependence

When diffusivity is concentration dependent, namely $D = D(m)$, the field (3.6) is modified to read

$$\frac{\partial m}{\partial t} = \frac{\partial}{\partial x}\left(D(m)\frac{\partial m}{\partial x}\right), \tag{5.12}$$

whereby it becomes non-linear.

Several methods for the solution of (5.12) were presented in the past (Crank 1980) and explored for specific assumed forms of $D(m)$ – mostly linear and exponential. In a more recent work (Lee and Peppas 1993), simulations for the latter case were evaluated numerically and exhibited in detail, together with the associated residual stresses. The characteristic features of concentration dependence were listed by Rogers (1965). Among those it is important to note that $M(t^*)$ vs. t^* depends on the specimen's thickness and that desorption varies significantly from absorption. While the foregoing features may serve to identify the presence of concentration dependence diffusivity, the actual task of expressing that dependence is formidable indeed.

5.5 Fiber Distribution and Shape

A detailed numerical study (Bond 2005) has shown that for the same overall fiber volume fraction, the diffusion coefficient D is rather insensitive to the spacing or distribution of circular fibers, although, in the case of composite laminates, the presence of a resin-rich interlaminar region would affect the overall value of the transverse component of D.

On the other hand, it was demonstrated by means of a thorough computational scheme (Aditya and Sinha 1996) that D is highly sensitive (up to 30%) to fiber shape and that this sensitivity increases with volume fraction.

Besides circular and elliptical fiber cross sections, that work investigated the effects of more complex, realistic shapes depicted in Fig. 5.3 below

5.6 Stress Assisted Diffusion

Upon the incorporation of stress effects on the free volume (Knauss and Emri 1981), namely

$$V_{\mathrm{f}} = V_{\mathrm{fo}} + \alpha\Delta T + \gamma m + \varepsilon_{\mathrm{kk}}^{\mathrm{f}}, \tag{5.13}$$

where $\varepsilon_{\mathrm{kk}}^{\mathrm{f}}$ denotes the dilatation of the free volume V_{f} due to mechanical stresses, it was possible to incorporate their effects on D within the form (Lefebvre 1987)

$$D = D_0\left(\frac{T}{T_0}\right)^2 \frac{\rho}{\rho_0} \exp\left(-\frac{E}{RT}\right) \exp\left(\frac{B}{V_{\mathrm{fo}}} \frac{\alpha\Delta T + \gamma m^N + \varepsilon_{\mathrm{kk}}^{\mathrm{f}}}{V_{\mathrm{fo}} + \alpha\Delta T + \gamma m^N + \varepsilon_{\mathrm{kk}}^{\mathrm{f}}}\right), \tag{5.14}$$

where ρ denotes density and the power N is added for generality. Subscripts "0" denote reference values.

While fiber spacings and distributions have but a small effect on diffusivity they play a major role in the residual hygral stresses within a fiber reinforce composite.

Computational investigations of the distribution of the residual octahedral shear stress in regularly arrayed fibrous composites (Miller and Adams 1978) as well as interfacial normal stresses (Lee and Peppas 1993) have shown that those stresses, which vary between tension and compression, may reach amplitudes of up to 30 MPa. Those investigations have also shown that the above stress distributions varied with time and fiber volume fraction.

The effect of nonuniform fiber distribution was shown (Tsotsis and Weitsman 1990) to introduce a stress concentration factor of 3, indicating that residual hygrothermal stresses are likely to exceed the strength of polymers at certain locations within the composite.

A different approach to evaluate the effects of stresses on D (Neumann and Marom 1987) related weight gain data to stresses that arise in composites oriented

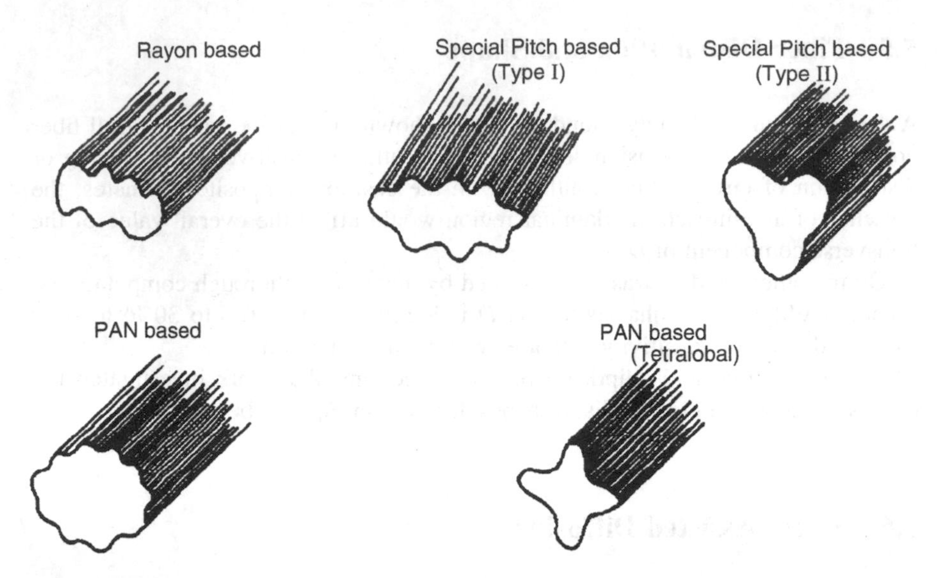

Fig. 5.3 Microstructural morphology of variously shaped carbon fiber. Reprinted from Aditaya and Sinha (1996), Copyright (1996), with permission from Elsevier

at several angles about the load direction. The effect of those stresses was incorporated into D by considering their contributions to the free volume, employing a concept somewhat akin to that expressed in (5.13).

They obtained excellent agreement between predicted and recorded values of $M(t)$ vs. $\sqrt{t}$ for various values of σ, as shown in Fig. 5.4 below. Their predictions for D were compared against data plotted vs. nondimensional combinations of parameters employed in their formulation.

In another approach, the deviation from Fickian diffusion is attributed to the dependence of the diffusion coefficient D on the volumetric strain ε within the matrix. (Youssef et al. 2009). Employing an elastic stress analysis and reasonable assumptions regarding the equilibrium moisture content for the unstrained state, a time and space iterative scheme was developed to compute the coupling effects between strain and moisture uptake.

A comprehensive formulation of stress-assisted diffusion (Weitsman 1987a) was established upon employing fundamental concepts of irreversible thermodynamics and continuum mechanics.

Accordingly, consider a solid body B occupying a material volume V bounded by a surface A. Let the solid, of mass density ρ_s, absorb vapor through its boundary and let m denote the vapor-mass per unit volume of the solid. Also, and let $\mathbf{f}$, $\mathbf{q}$, and $\mathbf{v}$ denote fluxes of vapor-mass and of heat, and the velocity of the solid particles, respectively.

In addition, let u and s be the internal energy and entropy densities of the solid/vapor mixture per unit solid mass and let σ_{ij} and T denote the components of stress due to mechanically applied loads and temperature, respectively.

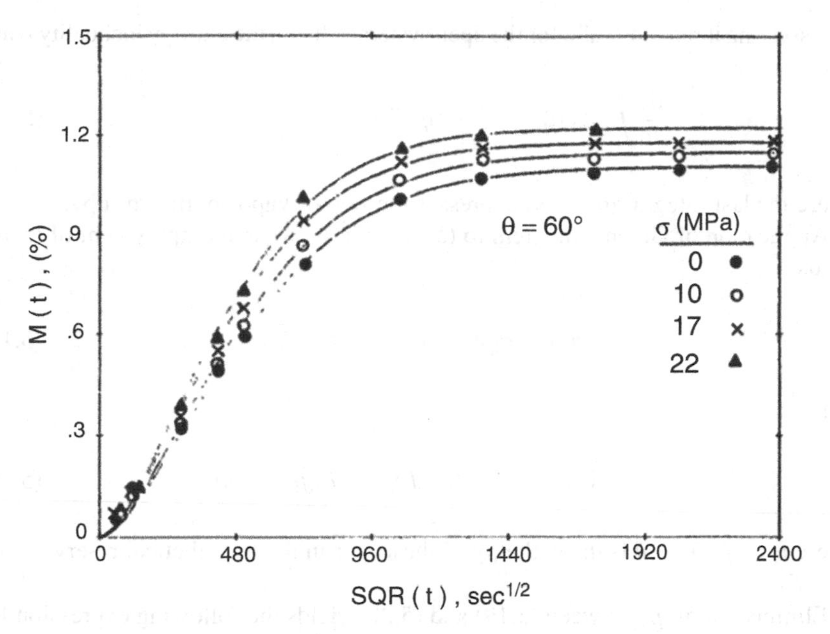

Fig. 5.4 The percent moisture content as a function of square root of exposure time for specimens of $\phi_f = 0.45$, loaded at $\theta = 60°$ by different stresses. *Dotted line* is calculated line based on Fickian solution (after Neumann and Marom 1987)
Credit: Neumann S and Marom G. "Prediction of moisture diffusion parameters in composite materials under stress." 21(1), pp 68–80, Copyright 1987 by *Journal of Composite Materials*. Reprinted by permission of SAGE

A proper accounting of the state of the solid/vapor mixture, which is a thermodynamically open system, is obtained by considering each element in thermodynamic equilibrium with a reservoir containing vapor at pressure $\tilde{p}$ density $\tilde{\rho}$, and internal energy and entropy densities $\tilde{u}$ and $\tilde{s}$, respectively.

Conservation of the solid and vapor masses gives

$$\dot{\rho}_s + \rho_s \nabla \cdot \mathbf{v} = 0, \tag{5.15}$$

$$\dot{m} = -\nabla \cdot \mathbf{f}. \tag{5.16}$$

Conservation of energy over B reads

$$\frac{\mathrm{d}}{\mathrm{d}t}\int_V \rho_s u \,\mathrm{d}V = \int_A \rho_{ij} n_j v_i \,\mathrm{d}A - \int_A q_i n_i \,\mathrm{d}A - \int_A \tilde{p}\frac{f_i}{\tilde{\rho}} n_i \,\mathrm{d}A - \int_A \tilde{u} f_i n_i \,\mathrm{d}A. \tag{5.17}$$

The third integral on the right side of (5.17) expresses the mechanical power due to vapor flux, observing that $f_i/\tilde{\rho}$ corresponds to vapor velocity. The last integral in (5.17) expresses the rate of vapor-borne energy.

Assuming it remains valid for the open system at hand, the entropy inequality reads

$$\frac{\mathrm{d}}{\mathrm{d}t}\int_V \rho_s s\,\mathrm{d}V \geq \int_A -(q_i/T)n_i\mathrm{d}A - \int_A \tilde{s}f_i n_i\mathrm{d}A, \tag{5.18}$$

where the last integral in (5.18) expresses the rate of vapor-borne entropy.

Application of Green's theorem to (5.17) and (5.18), and employment of (5.16) yields

$$\rho_s\dot{u} = \sigma_{ij}v_{i,j} - q_{i,i} + \tilde{h}_{,i}f_i + \tilde{h}\dot{m} \tag{5.19}$$

and

$$\rho_s T\dot{s} \geq -q_{i,i} + (q_i/T)g_i - T\tilde{s}_{,i}f_i + T\tilde{s}\dot{m}, \tag{5.20}$$

where $\tilde{h} = (\tilde{p}/\tilde{\rho}) + \tilde{u}$ is the enthalpy of the vapor in the hypothetical reservoir and $g_i = T_{,i}$.

Elimination of $q_{i,i}$ between (5.19) and (5.20) yields the following expression for the "reduced entropy inequality":

$$-\rho_s\dot{\psi} - \rho_s s\dot{T} + \sigma_{ij}v_{i,j} - (q_i/T)g_i + \tilde{\mu}\dot{m} - f_i\tilde{\mu}_{,i} - \tilde{s}g_i f_i \geq 0. \tag{5.21}$$

In (5.21), $\psi = u - Ts$ is the Helmholz free energy and $\tilde{\mu} = \tilde{h} - T\tilde{s}$ is the chemical potential of the vapor in the hypothetical reservoir.

Assuming isotropy, the linear elastic strain formulation is derived by setting $\psi = \psi(I_1^\epsilon, (I_1^\epsilon)^2, I_2^\epsilon, T, m)$ where $I_1^\epsilon = \epsilon_{kk}$ and I_2^ϵ are the first and second isotropic invariants of the infinitesimal strain ϵ_{ij}. Note also that since $\epsilon_{ij} \ll 1$, the density $\rho_s \cong \rho_{s0}$.

Expansion of ψ in powers in ε_{ij} gives

$$\rho_s\psi = A(m,T) - C(m,T)\varepsilon_{kk} + \frac{1}{2}\lambda(m,T)\varepsilon_{ii}\varepsilon_{jj} + G(m,T)\varepsilon_{ij}\varepsilon_{ij}. \tag{5.22}$$

Since $\sigma_{ij} = \rho_{s0}(\partial\psi/\partial\varepsilon_{ij})$ it follows that

$$\sigma_{ij} = -C(m,T)\delta_{kk} + \lambda(m,T)\varepsilon_{kk}\delta_{ij} + 2G(m,T)\varepsilon_{ij} \tag{5.23}$$

and

$$\tilde{\mu} = A_{,m}(m,T) - C_{,m}(m,T)\varepsilon_{kk}. \tag{5.24}$$

Omitting several intermediate steps (for details see Weitsman 1987a), and assuming that all gradients are sufficiently small so that their products may be neglected,[3] one obtains the following system of equations that couple the effects of strain, fluid mass, and temperature

$$\frac{\partial m}{\partial t} = D_{\mathrm{MM}}\nabla^2 m + D_{\mathrm{MT}}\nabla^2 T + D_{\mathrm{ME}}\nabla^2 \varepsilon \tag{5.25}$$

and

$$\rho_{s0} C_{\mathrm{EM}}\frac{\partial T}{\partial t} - C_{\mathrm{EE}}\frac{\partial \varepsilon}{\partial t} - C_{\mathrm{MM}}\frac{\partial m}{\partial t} = K_{\mathrm{TT}}\nabla^2 T + K_{\mathrm{TM}}\nabla^2 m + K_{\mathrm{TE}}\nabla^2 \varepsilon. \tag{5.26}$$

In addition, (5.23) together with $\sigma_{ij,j} = 0$ gives

$$- C_m m_{,i} - C_{,T} T_{,i} + (\lambda + G)\varepsilon_{,i} + G u_{i,\mathrm{kk}} + 0(\varepsilon) = 0. \tag{5.27}$$

In the above $\varepsilon = \varepsilon_{\mathrm{kk}}$.

The coefficients in (5.23) and (5.26) depend on m, T, and ε while those in (5.27) depend on m and T only.

Note that in the absence of moisture (5.26) and (5.27) reduce to those of coupled thermoelasticity.

When thermal effects are discarded, (5.24) and (5.27) reduce to

$$\frac{\partial m}{\partial t} = D_{\mathrm{MM}}\nabla^2 m + D_{\mathrm{ME}}\nabla^2 \varepsilon \tag{5.28}$$

and

$$- C_{,m} m_{,i} + (\lambda + G)\varepsilon_{,i} + G u_{i,\mathrm{kk}} = 0. \tag{5.29}$$

Equations (5.28) and (5.29) express the coupled strain-diffusion relations.

If all factors in the above equations are assumed constant, moisture diffusion uncouples from the strain field. This uncoupling is attained by taking the divergence of (5.29), whereby

$$C_{,m}\nabla^2 m + (\lambda + 2G)\nabla^2 \varepsilon = 0,$$

whereupon $\nabla^2 \varepsilon = (C_{,m}/(\lambda + 2G))\nabla^2 m$ can be substituted into (5.28) to give the *uncoupled* expression

$$\frac{\partial m}{\partial t} = D\nabla^2 m \quad \text{with} \quad D = D_{\mathrm{MM}} + \frac{c_{,m}}{\lambda + 2G} D_{\mathrm{ME}}. \tag{5.30}$$

[3] This implies that terms like $\nabla m \bullet \nabla m$ and alike are negligible relative to $\nabla^2 m (= D(\partial m/\partial t))$.

An analogous stress formulation is derivable in terms of the Gibbs free energy function $\phi(m,T\sigma_{ij})$, defined as

$$\rho_{s0}\phi = \rho_{s0}\psi - \sigma_{ij}\varepsilon_{ij}, \tag{5.31}$$

where

$$\varepsilon_{ij} = -\rho_{s0}\frac{\partial\phi}{\partial\sigma_{ij}}, \tag{5.32a}$$

$$s = -\frac{\partial\phi}{\partial T}, \tag{5.32b}$$

$$\tilde{\mu} = \rho_{s0}\frac{\partial\phi}{\partial m}. \tag{5.32c}$$

For sufficiently small stresses (relative to, say, a characteristic failure-stress) the expansion of ϕ in powers of σ_{ij} reads

$$-\rho_{so}\phi = -a(m,T) + B(m,T)\sigma_{\mathrm{kk}} + \frac{1}{2E}\sigma_{ii}\sigma_{jj} - \frac{1+v}{2E}(\sigma_{kk}\sigma - \sigma_{ij}\sigma_{ij}). \tag{5.33}$$

In (5.33), $E = E(m, T)$ and $v = v(m, T)$ denote Young's modulus and Poisson's ratio, respectively.

Denote $\sigma_{\mathrm{kk}} = \sigma$, employment of (5.32) and (5.33) then yields

$$\varepsilon_{ij} = B(m,T)\delta_{ij} + \frac{1+v}{E}\sigma_{ij} - \frac{v}{E}\sigma\delta_{ij}, \tag{5.34}$$

$$\tilde{\mu} = -a_{,m}(m,T) + B_m(m,T)\sigma + \frac{1}{2}\frac{\partial E}{\partial m}\left(\frac{\sigma}{E}\right)^2 + \frac{1}{2}\left(E\frac{\partial v}{\partial m} - (1+v)\frac{\partial E}{\partial m}\right) \times \frac{\sigma^2 - \sigma_{ij}\sigma_{ij}}{E^2}. \tag{5.35}$$

For most materials,

$$\left(\frac{\partial v}{\partial m}, \frac{1}{E}\frac{\partial E}{\partial m}, \frac{\sigma_{ij}}{E}\right) \ll 1.$$

Consequently (5.35) reduces to

$$\tilde{\mu} = a_{,m}(m,T) - B_{,m}(m,T)\sigma. \tag{5.36}$$

If all gradients $\nabla\sigma$, ∇m, and ∇T are sufficiently small, their products may be discarded. Then, upon dispensing with several intermediate steps one obtains the coupled stress-diffusion expression

$$\frac{\partial m}{\partial t} = \hat{D}_{\mathrm{MM}}\nabla^2 m + \hat{D}_{\mathrm{M}\sigma}\nabla^2\left(\frac{\sigma}{\sigma_0}\right). \tag{5.37}$$

Recall the compatibility condition

$$\varepsilon_{ij,kl} + \varepsilon_{kl,ij} - \varepsilon_{ik,jl} - \varepsilon_{jl,ik} = 0. \tag{5.38}$$

Setting $i = k$ and $j = l$, summing, employing (5.38), and recalling $\sigma_{ij,j} = 0$, it follows that

$$\nabla^2\sigma + \frac{2E}{1-\nu}\beta\nabla^2 m = 0. \tag{5.39}$$

It can be readily observed that the elimination of $\nabla^2\sigma$ between (5.37) and (5.39) results in an uncoupled moisture diffusion equation

$$\frac{\partial m}{\partial t} = \hat{D}\nabla^2 m, \tag{5.40}$$

with

$$\hat{D} = \hat{D}_{\mathrm{MM}} - 2\frac{E\beta}{(1-\nu)\sigma_0}\hat{D}_{\mathrm{M}\sigma}. \tag{5.41}$$

In analogy with (5.30), the diffusion process still can be stress-dependent when $\hat{D}_{\mathrm{MM}}$, $\hat{D}_{\mathrm{M}\sigma}$, and thereby $\hat{D}$ depend on stress. Obviously these parameters are also temperature sensitive.

It is worth noting that upon accounting for the molecular weight M_{w} of water, it was possible to refine (5.36) to read (Derrien and Gilormini 2007)

$$B_{,m}(m, T_0) = \frac{3M_{\mathrm{w}}\beta}{\rho_{\mathrm{p}}}, \tag{5.42}$$

where ρ_{p} denotes the specific density of the polymer.

A somewhat different result, where $B_{,m} = \beta/\rho$ was obtained elsewhere (Wu 2001). Employing Eshelby's stress (Eshelby 1951) that article presented a scheme that resulted in coupled stress-diffusion formulations as well.

5.7 Diffusion Coupled with Damage

The incorporation of damage variables within the context of continuum mechanics implies the presence of a recurrent distribution of microcracks or microvoids within a sufficiently small volume element that represents the entire material.

In any such continuum formulation the magnitude of the representative volume element should be much smaller than any structural dimension of the material volume at hand. For this reason, the following formulation applies to the early stage of fluid-induced damage prior to the coalescence of the individual arc-shaped microcracks to form the long hairline cracks at the fiber matrix interfaces shown in Fig. 4.30a–c.

The basic formulation follows (5.15)–(5.21), with the presence of damage introduced by incorporating it within the Helmholz free energy ψ, namely (Weitsman 1987b)

$$\psi = \psi(\varepsilon_{ij}, T, m, d_{[ij]}), \tag{5.43}$$

where the skew symmetric tensor $d_{[i\ j]}$ accounts for the totality of the three components of the face areas of the microcracks within a representative volume element. The choice of skew symmetry allows for the fact that each microcrack is contained within two equal, but of opposite sign, surfaces.

Considering transverse isotropy, as applies to unidirectionally fiber reinforced composites, it is necessary to express (5.43) in terms of the individual and joint invariants of ε_{ij} and $d_{[i\ j]}$ for that particular symmetry. Such an expansion yields an expression for ψ that involves 17 material parameters, all of which are functions of m, T, $d^2_{[12]}$ and $d^2_{[31]} + d^2_{[23]}$. Obviously, such a general formulation is practically intractable.

A simplification occurs for the realistic isothermal case of diffusion across the thickness direction x_1 of a plate, where all fibers run parallel to x_2. In this case ψ contains eight parameters.

Reverting to a stress formulation in terms of the Gibbs free energy function ϕ it follows that there remain only two components of damage $d_{[31]}$ and $d_{[23]}$ and that for sufficiently small values of σ_0 (i.e., relative to the failure of stress σ_f), their growth depends at most linearly on σ_0.

Furthermore, in this case, moisture flux has two components f_1 and f_2, both of which are related to a single component of the gradient of chemical potential $z_1 = \partial\tilde{\mu}/\partial x_1$, according to

$$f_1 = [A_1 + A_2(d^2_{([31])} - d^2_{([23])})]z_1 \tag{5.44}$$

and

$$f_2 = 2A_2 d_{[31]} d_{[32]} z_1 \tag{5.45}$$

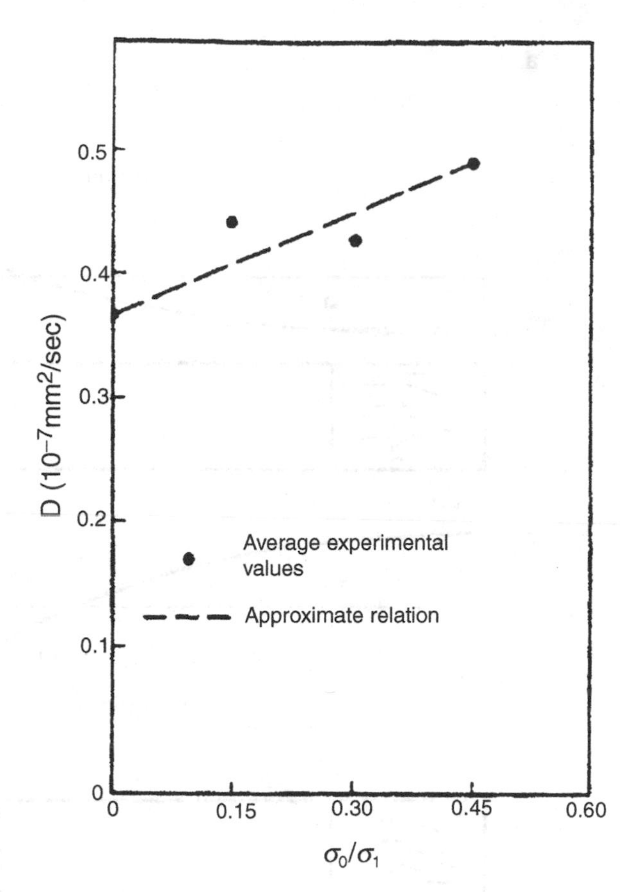

Fig. 5.5 Variation of the "equivalent Fickian" moisture-diffusivity coefficient D with stress (Weitsman 1987b)

Although the above approach yielded no expression for the diffusivity D, the aforementioned linear dependence on σ_0/σ_f may be reflected in experimental data that can be correlated with those parameters as shown in Fig. 5.5 above.

A micromechanical correlation between damage and diffusion was formulated on the basis of fluid-induced residual stresses (Guo and Weitsman 2002).

Consider an extended unidirectionally reinforced plate of thickness $2l$. Let x denote the thickness coordinate and consider all fibers to run parallel to the coordinate z.

The self-evident requirement that fluid-induced residual stresses should be self-equilibrating yields the expression

$$\sigma(x,t) = \sigma_0 - E\beta\left[m(x,t) - \frac{1}{2l}\int_{-l}^{l} m(x,t)\,\mathrm{d}x\right], \tag{5.46}$$

where E and β denote transverse modulus and fluid expansion coefficient, respectively, σ_0 is the in-plane stress acting in the y-direction and $m(x;t)$ are fluid distribution profiles as predicted, say, by Fick's law (3.8) and (3.9).

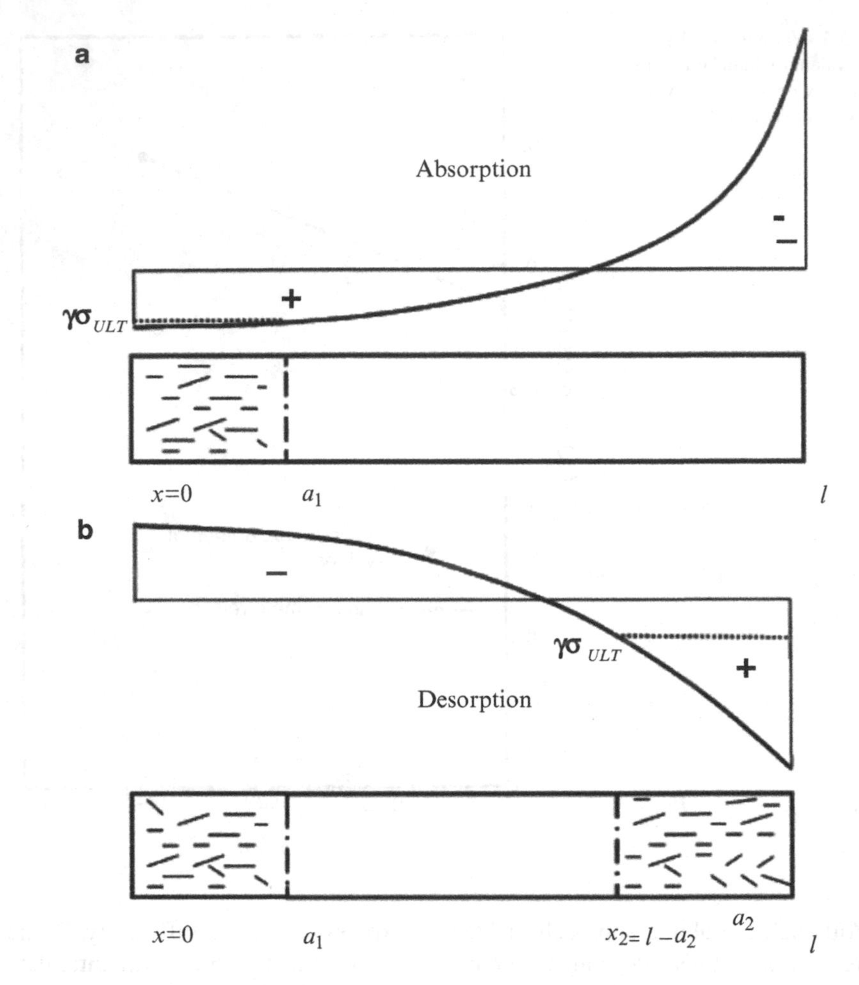

Fig. 5.6 Typical distribution of hygrothermal stresses during absorption and desorption processes and the corresponding damage regions (Weitsman and Guo 2002)

Damage in the form of microcracks is considered to occur within regions where tensile stresses $\sigma(x,t)$ satisfy the condition

$$\gamma\sigma(x,t) \geq \sigma_{\text{ult}} \tag{5.47}$$

In the above, γ denotes a stress concentration factor that arises from the characteristic fiber reinforced geometry as well as due to the nonuniformities in fiber spacings. Typically, one may consider $\gamma \sim 3$.

In the absence of applied stresses, the ingress of fluid across the boundaries $|x| = l$ would result in tensile stresses over the interior region $|x| \leq a_1$ during absorption, and within the exterior regions $l - a_2 \leq |x| \leq l$ during desorption as shown in Fig. 5.6 (where, due to symmetry, only half of the composite layer is shown).

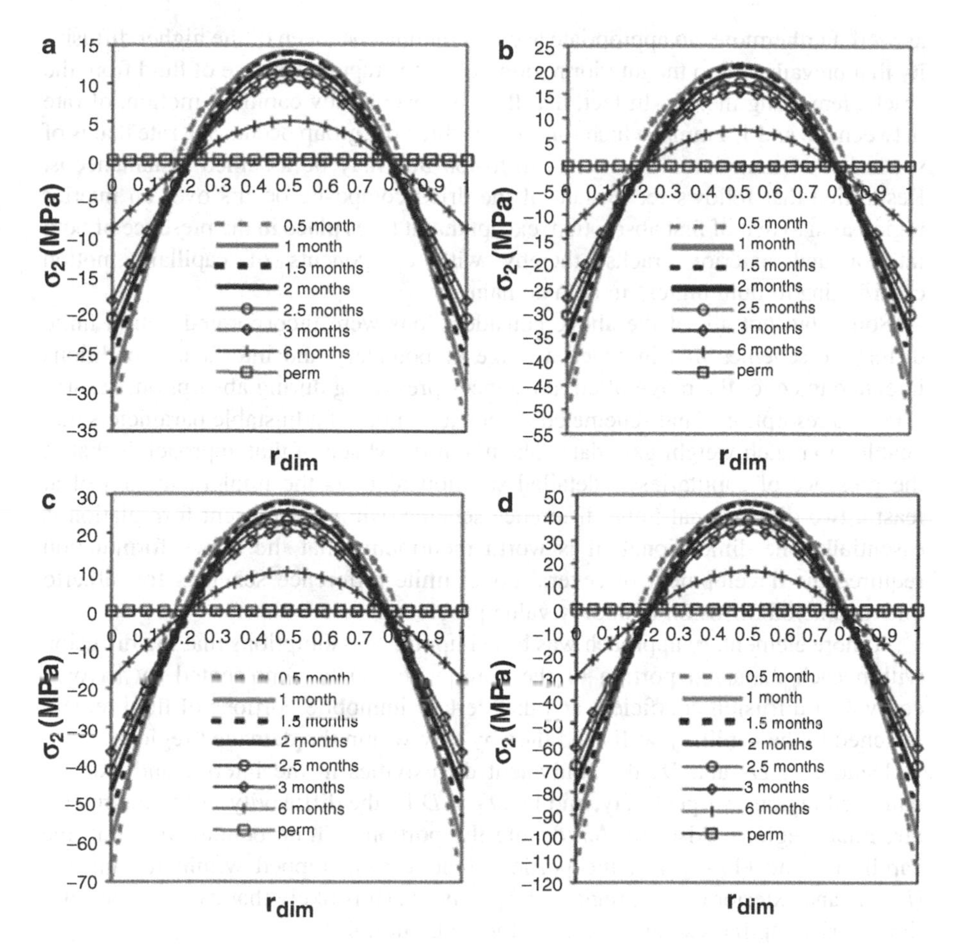

Fig. 5.7 T300/5208 transverse macroscopic stresses for several moisture content ratios (Jacquemin et al. 2005)
Credit: Jacquemin F, Freour S, Guillen R. "A hygroelastic self-consistent model for fiber-reinforced composites." 24(5), pp 485–502, Copyright 2005 by *Journal of Reinforced Plastics and Composites*, Reprinted by permission of SAGE

Although both a_1 and a_2 vary with time, this temporal variation is rather small, with typical results demonstrated in Fig. 5.7 for a circular region (Jacquemin et al. 2005).

During the absorption stage, internal cracks are assumed to form within the interior region, with lengths not exceeding $\hat{a}_1$ that correspond to the extent of the zone over which the condition (5.47) prevails. The spacing of those interior cracks can be estimated by means of fracture energy considerations akin to those employed for thermally induced fracture in cross ply laminates (Fang et al. 1989). An analogous analysis can be performed for the exterior cracks under desorption, except that, at this stage, the presence of the interior cracks must be considered as well. Furthermore, an appropriate accounting must be taken of the higher diffusivity that prevails within the interior region due to the rapid discharge of fluid from the cracks remaining therein. In fact, that flow is governed by capillary motion, of rate between 1/6 and 1/4 mm/s which for a 1-mm thick ply group occurs at a rate that is of six orders of magnitude faster than diffusion and may be assumed instantaneous. Resorption that follows reexposure of the dried composite occurs over an interior region akin to that of first absorption, except that it takes place in the presence of both interior and exterior cracks, thereby with components of capillary motion contributing to fluid ingress in both domains.

Some, but not all, of the above considerations were incorporated within finite difference schemes that included a range of boundary and interfacial conditions intended to cover the range of circumstances prevailing during absorption, desorption, and resorption. That scheme contained a number of adjustable parameters that enabled to match weight gain data. The major drawback of that approach is that in the presence of capillaries, a detailed solution requires the implementation of at least a two-dimensional finite difference scheme, while the present formulation is essentially one dimensional. It is worth mentioning that the above formulation required the development of several novel finite difference schemes for hitherto unencountered diffusion boundary value problems.

A more elementary approach was based upon the assumptions that the diffusion within each damaged portion of the composite can be represented by its own equivalent diffusion coefficient and that certain immobile portions of fluid remain attached to the capillary walls and thereby stay within the damaged regions.

Denote by D_1 and D_3 the equivalent diffusivities in the interior and exterior damaged regions, respectively, and let $D_2 = D$ be the diffusivity of the intermediate, intact regions. Also, let Δm denote the portion of fluid contained within the capillaries and $(1 - \xi_i)\Delta m$ the portions that remain trapped within the interior $(i = 1)$ and exterior $(i = 2)$ regions, respectively. (It is likely that $\xi_2 < \xi_1$, because the exterior capillaries have access to the ambient fluid).

Assuming Fickian diffusion, the absorption process is governed by the following expressions

$$\frac{\partial m}{\partial t} = D_2 \frac{\partial^2 m}{\partial x^2} \quad a_1 < |x| < l \text{ for the intact region} \tag{5.48a}$$

and

$$\frac{\partial m}{\partial t} = D_1 \frac{\partial^2 m}{\partial x^2} \quad 0 < |x| < a_1 \text{ for the inner damaged zone region.} \tag{5.48b}$$

The interface conditions at $x = a_1$ read

$$D_1 \left. \frac{\partial^2 m}{\partial x^2} \right|_{a_1^-} = D_2 \left. \frac{\partial^2 m}{\partial x^2} \right|_{a_1^+} \tag{5.49a}$$

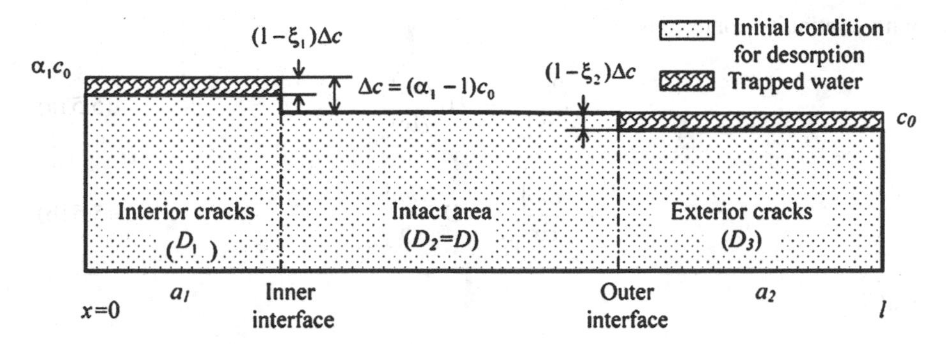

Fig. 5.8 Distribution of fluid concentration at the fully-saturated stage of absorption and initial condition for desorption (Weitsman and Guo 2002)

and

$$m_1(a_1^-) = \alpha_1 m(a_1^+). \tag{5.49b}$$

Expressions (5.49a, b) are identical with (3.18) and (3.19). Finally, due to symmetry, it follows that

$$\left.\frac{\partial m}{\partial x}\right|_{x=0} = 0 \tag{5.49c}$$

Rather than employing the analytical form detailed in Sect. 3.4, the solution for the present case was generated numerically.

The equilibrium distribution at saturation, with $m(x,\infty) = m_0$ for $a_1 < |x| < l$ and $m(x, \infty) = a_1\, m_0$ for $0 < |x| < a_1$ is shown in Fig. 5.8.

In view of the presence of entrapped portions within damaged regions, the above figure depicts the initial conditions for desorption from the initially saturated state.[4]

The desorption process is governed by the following expressions:

$$\frac{\partial m}{\partial t} = D_1 \frac{\partial^2 m}{\partial x^2} \quad 0<|x|<a_1 \text{ for the intact region,} \tag{5.50a}$$

$$\frac{\partial m}{\partial t} = D_2 \frac{\partial^2 m}{\partial x^2} \quad \alpha<|x|<l-a_2 \text{ for the intact region.} \tag{5.50b}$$

$$\frac{\partial m}{\partial t} = D_3 \frac{\partial^2 m}{\partial x^2} \quad l-a_2<|x|<l \text{ for the intact region} \tag{5.50c}$$

[4] Capillary motion is assumed to occur instantly upon the introduction of damage in the exterior region, since it moves 5–6 orders of magnitude faster than diffusion.

with continuity conditions

$$D_1 \left.\frac{\partial m}{\partial x}\right|_{a_1^-} = D_2 \left.\frac{\partial m}{\partial x}\right|_{a_1^+}, \tag{5.51a}$$

$$D_2 \left.\frac{\partial m}{\partial x}\right|_{a_2^-} = D_3 \left.\frac{\partial m}{\partial x}\right|_{a_2^+} \tag{5.51b}$$

and

$$m_1(a_1^-) = a_1^1 m(a_1^+), \tag{5.51c}$$

$$m_1(a_2^-) = m(a_2^+)/a_3^1, \tag{5.51d}$$

where a_1^1 and a_3^1 are related to a_1 and a_3 upon eliminating the entrapped fluids from the diffusing portions.

Condition (5.49c) holds as well.

The solution was generated numerically.

The formulation for resorption resembles that of desorption, except that the initial conditions incorporated the presence of the entrapped portions of fluid within the damaged zones, namely

$$m(x,0) = \begin{cases} (1-\xi_1)\Delta m & 0<|x|<a_1, \\ 0 & a_1<|x|<l-a_2, \\ (1-\xi_2)\Delta m & l-a_2<|x|<l. \end{cases} \tag{5.52}$$

Results are shown in Figs. 5.9–5.11

Note the discontinuities in moisture content at the boundaries of the various damaged regions. Similar abrupt variations occur for the more detailed capillary model as well.

The predicted weight gains (or weight loss) vs. $\sqrt{t}$ plots are shown for both models in Fig. 5.12 for the parameters listed in that article.

It was demonstrated by means of a parametric study that the effect of D_1 on the predictions of weight gain/loss/regain is negligible, and the results are strictly dominated by the choice of D_3. It is therefore possible to calibrate the values of the three parameters D_3, $\xi_1 \Delta m$, and $\xi_2 \Delta m$ in order to fit absorption–desorption–resorption data. Of the above, the latter two correspond to the gaps between the saturation values.

A more comprehensive, though somewhat less detailed, micromechanical formulation of damage-affected diffusion considered a homogenized material consisting of a solid phase that contains distributed air channels (Lundgren and Gudmundson 1999). The moisture content and diffusivities in both phases are denoted by $m^{(n)}$ and $D_{ij}^{(n)}$, $n = 1{,}2$ for the solid and channel phases, respectively.

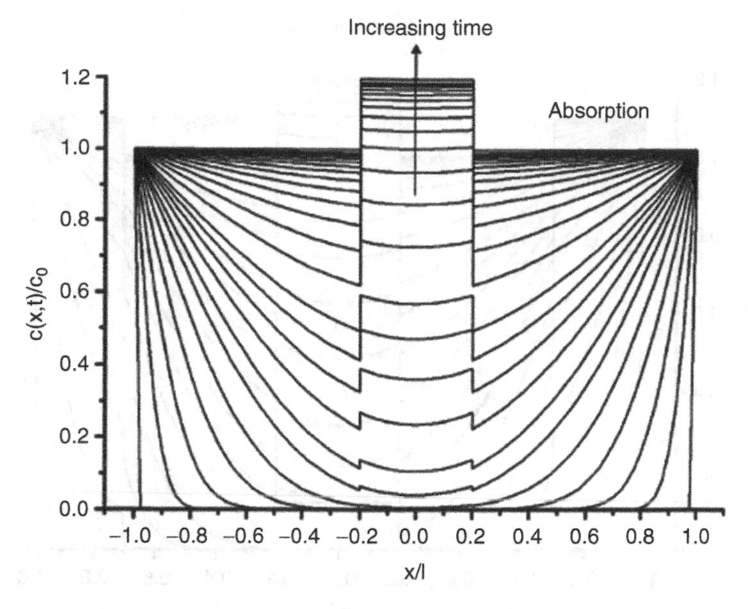

Fig. 5.9 Fluid concentration profiles at various times during absorption in the presence of an interior damage region, calculated by the equivalent-diffusivity model (Weitsman and Guo 2002)

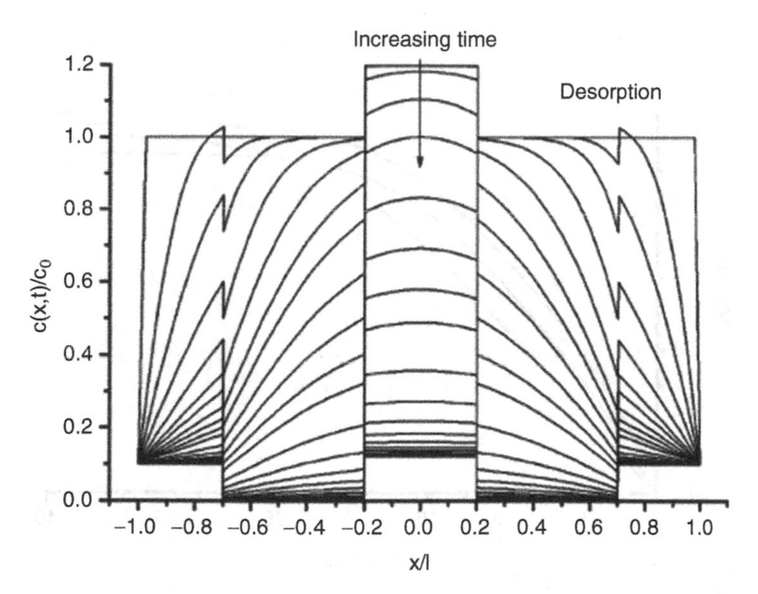

Fig. 5.10 Fluid concentration profiles at various times during desorption, in the presence of an interior- and exterior-damage regions, calculated by the equivalent-diffusivity model (Weitsman and Guo 2002)

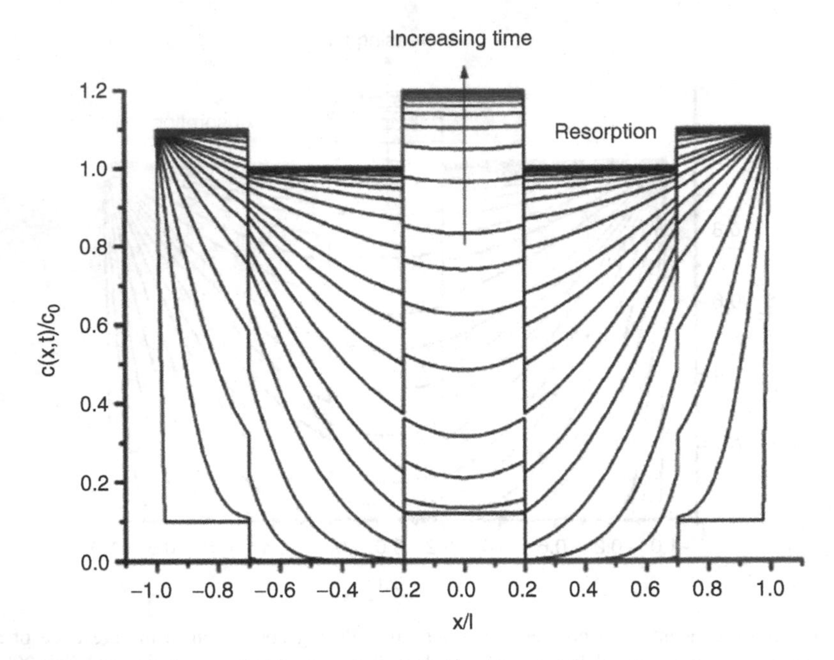

Fig. 5.11 Fluid concentration profiles at various times during resorption, in the presence of an interior- and exterior-damage regions, calculated by the equivalent-diffusivity model (Weitsman and Guo 2002)

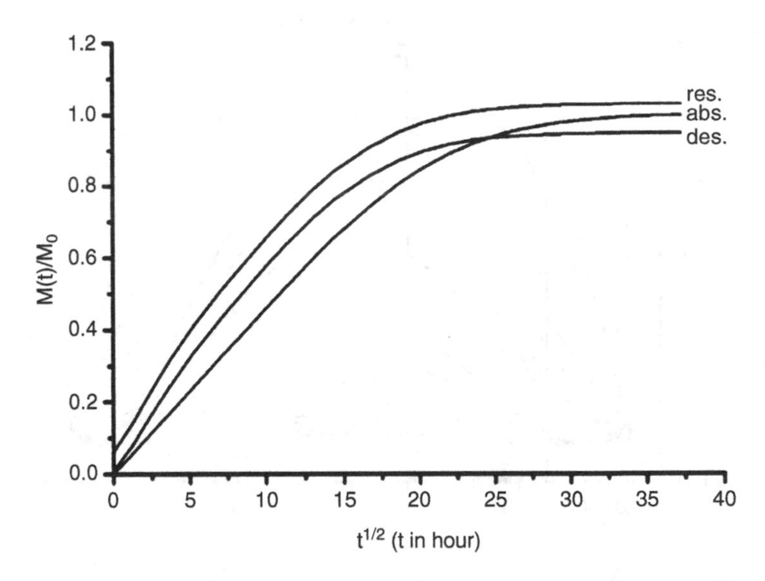

Fig. 5.12 Fluid weight gain (or weight-loss) curves for initial sorption, desorption and resorption processes, as calculated by the equivalent diffusivity model, plotted vs. $\sqrt{time}$ (Weitsman and Guo 2002)

At equilibrium one has $m_\infty^{(1)} = km_\infty^{(2)}$. Thus, at all prior times it is assumed that the interphase flux q_{21} is given by

$$q_{21} = A(km^{(2)} - m^{(1)}), \tag{5.53}$$

where A represents a temperature-dependent inverse time factor.

The governing diffusion equations for $m^{(n)}$ $n = 1,2$ thus read

$$\dot{m}^{(n)} = D_{ij}^{(n)} m_{,ij}^{(n)} \pm A(km^{(2)} - m^{(1)}), \quad n = \begin{Bmatrix} 1 \\ 2 \end{Bmatrix}. \tag{5.54}$$

It was now possible to explore some specific limiting cases. If all channels are interconnected, the flow therein was much faster than the diffusion within the solid and $m^{(2)}$ approaches $m_\infty^{(2)}$ nearly instantaneously[5] and the problem reduces to that of diffusion within the solid phase alone, thus

$$\frac{\partial m^{(1)}}{\partial t} = D_\mathrm{T} \frac{\partial^2 m^{(1)}}{\partial x^2} + A(m_\infty^{(1)} - m^{(1)}). \tag{5.55}$$

In (5.55) D_T denotes diffusivity transverse to fiber direction.

If, in addition, the flux coefficient A is much larger than the time scale for diffusion $L^2 D_\mathrm{T}$, (5.55) is further simplified to read

$$\frac{\partial m^{(1)}}{\partial t} = A(m_\infty^{(1)} - m^{(1)}). \tag{5.56}$$

At the other extreme of extremely slow flux, i.e., $A \ll L^2 D_\mathrm{T}$, (5.55) reduces to that of classical diffusion.

If the channels are not interconnected, it was argued (Lundgren and Gudmundson 1999) that it is possible to neglect $D_{ij}^{(2)}$ in comparison to $D_{ij}^{(1)}$ and thereby expressions (5.54) reduce to those of two-phase diffusion, namely (3.28a, b). This supposition may be akin to the consideration that such channels attract fluid by means of capillary action (Guo and Weitsman 2002) that occurs so much faster than diffusion, to be conceived as instantaneous. Consequently, the total moisture uptake would consist of the amount diffused within the solid together with portion contained within the channels.

The above models fail to account for moisture-affected damage growth criteria, though limited attention to that issue was provided by correlating damage with residual stresses during exposure to cyclic humidity (Guo and Weitsman 2002)

[5] This is equivalent to the assumption of instantaneous saturation within the fractured regions in the aforementioned article (Guo and Weitsman 2002).

Another approach employing concepts of continuum damage mechanics (e.g., Lemaitre and Chaboche 1994) was presented by Perreux and Suri (1997). Studying the effects of cyclic fatigue on $[\pm 55°]_3$ glass fiber/poly epoxy composite pipes they noticed that in the absence of stress moisture uptake data could be modeled as a process of a two-phase diffusion (see Sect. 3.6). Damage in the axial direction was related to the decrease in the axial stiffness, namely $d_{zz} = \Delta E_{zz}/E_{zz}$ and its effect on diffusion was incorporated within a reduced time parameter $\tilde{t}$, assumed to be expressed by[6]

$$\tilde{t} = \frac{t}{(1 - d_{zz})}(1 - d_{zz}). \tag{5.57}$$

Collecting stress-strain data for cyclic fatigue in the axial direction and recording the reduction in stiffness in both axial and circumferential directions it was possible to relate the effects of damage to the equivalent moduli $\tilde{\underline{\underline{A}}}$ of the pipe to their undamaged values $\underline{\underline{A}}$ by

$$\tilde{\underline{\underline{A}}} = \underline{\underline{A}} + \underline{\underline{A}}_1 d_{zz} + \underline{\underline{A}}_{22} d_{zz}^2 \tag{5.58}$$

The determination of damage growth $\dot{d}_{zz}$ follows a scheme based upon the premises of continuum damage mechanics. This approach requires the incorporation of several internal variables, a damage surface, associative or nonassociative damage growth relations, and, in particular, a work hardening parameter R_D that has been assumed to depend on two-fluid content-dependent parameters α_D and P_D.

The above work demonstrates the complexity of the coupling of damage and diffusion.

5.8 Contradictory Effects of Hydrostatic Pressure

Unidirectionally reinforced $[0°_4]$AS4/3501-6 graphite/epoxy coupons were immersed in simulated sea water at room temperature within pressure chambers at 13.8 and 20.7 MPa for 2 months and weight gain data recorded periodically up to saturation (Gao and Weitsman 1998). A third set of specimens was immersed at atmospheric pressure 0.1 MPa. Each circumstance involved at least four replicate samples. All weight data could be matched with linear Fickian predictions. Results are summarized in Table 5.1.

As can be noted from Table 5.1, lower weight gain levels and larger data scatter were recorded under $p = 13.8$ MPa at either 0.1 or 20.7 MPa.

[6] While it may be more appealing to incorporate damage within the diffusion coefficients D, β and γ, its insertion within a common reduced time $\tilde{t}$ is technically advantageous. However, for simpler diffusion phenomena it may better to contain damage within D.

Table 5.1 Range of saturation levels and their average values at various levels of hydrostatic pressure p

p MPa (ksi)	Range of m_∞	$\bar{m}_\infty$ (%)
0.1 (0)	$1.9\% < m_\infty < 2.1\%$	2.0
13.8 (2)	$1.45\% < m_\infty < 1.9\%$	1.7
20.7 (3)	$1.7\% < m_\infty < 1.8\%$	1.8

This apparent paradox can be explained by the two contradictory effects of hydrostatic pressure predicted earlier. On one hand, the free volume V_f diminishes under compressive strain (5.13), thus decreasing the ingress of fluids into the polymer. On the other hand, external pressure enhances the chemical potential μ of the polymer (5.36) and (5.42), which increases the driving force for fluid ingress. These two effects appear to compete against each other at unequal degrees.

References

Aditya PK, Sinha PK (1996) Moisture diffusion in variously shaped fibre reinforced composites. Comput Struct 59(1):157–166

Berens A (1977) Diffusion and relaxation in glassy polymer powders: 1. Fickian diffusion of vinyl chloride in poly(vinyl choride). Polymer 18(7):697–704

Berens A, Hopfenberg H (1978) Diffusion and relaxation in glassy polymer powders: 2. Separation of diffusion and relaxation parameters. Polymer 19(5):489–496

Bond DA (2005) Moisture diffusion in a fiber-reinforced composite: part I – non-Fickian transport and the effect of fiber spatial distribution. J Compos Mater 39(23):2113–2141

Cai LW, Weitsman Y (1994) Non-Fickian moisture diffusion in polymeric composites. J Compos Mater 28(2):130–154

Carter HG, Kibler KG (1978) Langmuir-type model for anomalous moisture diffusion in composite resins. J Compos Mater 12(2):118–131

Crank J (1980) The mathematics of diffusion, 2nd edn. Oxford University Press, Oxford

De Wilde W, Frolkovic P (1994) The modelling of moisture absorption in epoxies: effects at the boundaries. Composites 25(2):119–127

De Wilde WP, Shopov PJ (1994) A simple model for moisture sorption in epoxies with sigmoidal and two-stage sorption effects. Compos Struct 27(3):243–252

Derrien K, Gilormini P (2007) The effect of applied stresses on the equilibrium moisture content in polymers. Scr Mater 56(4):297–299

Doolittle AK, Doolittle DB (1957) Studies in Newtonian flow. V. Further verification of the free-space viscosity equation. J Appl Phys 28(8):901

Eshelby JD (1951) The force on an elastic singularity. Philos Trans R Soc Lond A Math Phys Sci 244(877):87–112

Fang G-P (1986) Moisture and temperature effects in composite materials. Texas A&M University Report MM-5022-86-21, November 1986

Fang G-P, Schapery RA, Weitsman Y (1989) Thermally-induced fracture in composites. Engineering Fracture Mechanics 33(4):619–632

Gao J, Weitsman YJ (1998) The tensile mechanical properties and failure behavior of stitched T300 mat/urethane 420 IMR composite. The University of Tennessee Report MAES98-2.0-CM

Guo Y, Weitsman YJ (2002) Solution method for beams on nonuniform elastic foundations. J Eng Mech 128(5):592–594

Jackle J, Frisch HL (1986) Properties of a generalized diffusion equation with a memory. J Chem Phys 85(3):1621–1627
Jacquemin F, Freour S, Guillen R (2005) A hygroelastic self-consistent model for fiber-reinforced composites. J Reinforced Plast Compos 24(5):485–502
Knauss W, Emri I (1981) Non-linear viscoelasticity based on free volume consideration. Comput Struct 13(1–3):123–128
Lee MC, Peppas NA (1993) Water transport in graphite/epoxy composites. J Appl Polym Sci 47 (8):1349–1359
Lefebvre DR (1987) A diffusion coefficient for moisture uptake in adhesive: effect of temperature, stress and moisture concentration. Progress report CAS/ESM-87-7, Virginia Tech Center for Adhesive Science
Lemaitre J, Chaboche J (1994) Mechanics of solid materials. Cambridge University Press, Cambridge
Long FA, Richman D (1960) Concentration gradients for diffusion of vapors in glassy polymers and their relation to time dependent diffusion phenomena. J Am Chem Soc 82(3):513–519
Lundgren J, Gudmundson P (1999) Moisture absorption in glass-fibre/epoxy laminates with transverse matrix cracks. Compos Sci Technol 59(13):1983–1991
Macedo PB, Litovitz TA (1965) On the relative roles of free volume and activation energy in the viscosity of liquids. J Chem Phys 42(1):245
Miller AK, Adams DF (1978) Inelastic micromechanical analysis of graphite/epoxy composites subjected to hygrothermal cycling. In: Vinson JR (ed) Advanced composite materials-environmental effects, ASTM STP 658, American Society of Testing and Materials, pp 121–142
Neumann S, Marom G (1987) Prediction of moisture diffusion parameters in composite materials under stress. J Compos Mater 21(1):68–80
Perreux D, Suri C (1997) A study of the coupling between the phenomena of water absorption and damage in glass/epoxy composite pipes. Compos Sci Technol 57(9–10):1403–1413
Rogers CE (1965) Solubility and diffusivity. In: Fox D, Labes MM, Weissberger A (eds) Chemistry and physics of the organic solid state, vol 2. Wiley, New York
Roy S, Xu WX, Park SJ, Liechti KM (2000) Anomalous moisture diffusion in viscoelastic polymers: modeling and testing. J Appl Mech 67(2):391–396
Shopov PJ, Frolkovic P, De Wilde W (1996) Free and bond water type models of penetrant sorption in epoxies. Sci Eng Compos Mater 5(1):39–55
Tsotsis T, Weitsman YJ (1990) Energy release rates for cracks caused by moisture absorption in graphite/epoxy composites. J Compos Mater 24(5):483–496
Turnbull D, Cohen MH (1961) Free-volume model of the amorphous phase: glass transition. J Chem Phys 34(1):120
Vieth WR, Sladek KJ (1965) A model for diffusion in a glassy polymer. J Colloid Sci 20 (9):1014–1033
Vieth W, Howell J, Hsieh J (1976) Dual sorption theory. J Memb Sci 1:177–220
Weitsman YJ (1987a) Stress assisted diffusion in elastic and viscoelastic materials. J Mech Phys Solids 35(1):73–94
Weitsman YJ (1987b) Coupled damage and moisture-transport in fiber-reinforced, polymeric composites. Int J Solids Struct 23(7):1003–1025
Weitsman YJ, Guo Y (2002) A correlation between fluid-induced damage and anomalous fluid sorption in polymeric composites. Compos Sci Technol 62(6):889–908
Williams ML, Landel RF, Ferry JD (1955) The temperature dependence of relaxation mechanisms in amorphous polymers and other glass-forming liquids. J Am Chem Soc 77(14):3701–3707
Wu CH (2001) The role of Eshelby stress in composition-generated and stress-assisted diffusion. J Mech Phys Solids 49(8):1771–1794
Youssef G, Freour S, Jacquemin F (2009) Stress-dependent moisture diffusion in composite materials. J Compos Mater 43(15):1621–1637

Chapter 6
Hygrothermal Viscoelastic Response

6.1 Data and Time-Temperature-Moisture Shifts

While the effects of moisture on the mechanical response of materials as different as wool and wood were recognized for a long time, their incorporation within the theory of viscoelasticity was not implemented until the 1970s.

Perhaps the earliest work in this vein concerned the time-moisture effects on the relaxation of PVA and nylon 6 films (Onogi et al. 1962). Relaxation data collected at distinct levels of relative humidity were coalesced through horizontal shifts to form a "master relaxation curve" as shown in Figs. 6.1 and 6.2. The horizontal shift factor function a_{H} is plotted vs. RH in Fig. 6.3.

Extensive research in this area was conducted at the Institute of Polymer Mechanics in Riga, where creep data for polyester resin were collected at several levels of temperature and moisture content (Maksimov et al. 1972). The data were subsequently plotted vs. ln t (t is time) (Maksimov et al. 1975) as shown in Fig. 6.4a and were subsequently shifted horizontally parallel to the ln t axis so as to form the "best" continuous curve (Fig. 6.4b). This resulted in a three-dimensional plot of a time-temperature-moisture shift factor plot of ln a_{Tm} vs. T and m. Since in the earlier work (1972), the time-moisture shift factor a_m was considered to be independent of T, the combined factor was represented by a product, namely $a_{Tm}(T,m) = a_T(T) \cdot a_m(m)$.

A more careful consideration of the multiple creep data (Maksimov et al. 1975) yielded an interactive expression for a_{Tm}, namely

$$\ln a_{Tm}(T,m) = A_1 T + A_2^2 T^2 + A_{\mathrm{M}} m + A_4^2 M^2 + A_5 Tm, \qquad (6.1)$$

where $A_1 = 0.115/^{\circ}\mathrm{C}$, $A_2 = 0.0645/1\circ\mathrm{C}$, $A_3 = 2.918/1\%m$, $A_4 = 0.727/1\%m$, and $A_5 = 0.0197/(1^{\circ}\mathrm{C} \cdot 1\%m)$.

Y.J. Weitsman, *Fluid Effects in Polymers and Polymeric Composites*, Mechanical Engineering Series, DOI 10.1007/978-1-4614-1059-1_6,

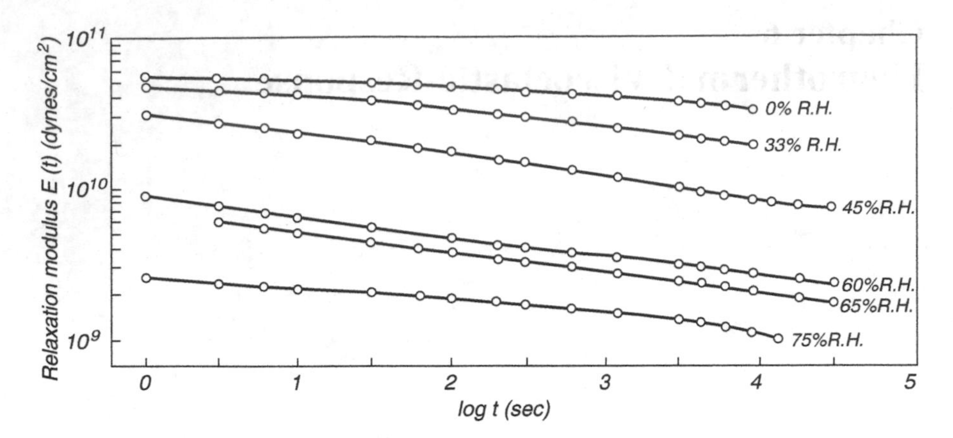

Fig. 6.1 Horizontal shift factors at various levels of relative humidity
Copyright (Onogi S, Sasaguri K, Adachi T, Ogihara S (1962). "Time-humidity superposition in some crystalline polymers." *Journal of Polymer Science*, 58(166): 1–17) "This material is reproduced with permission of John Wiley & Sons, Inc."

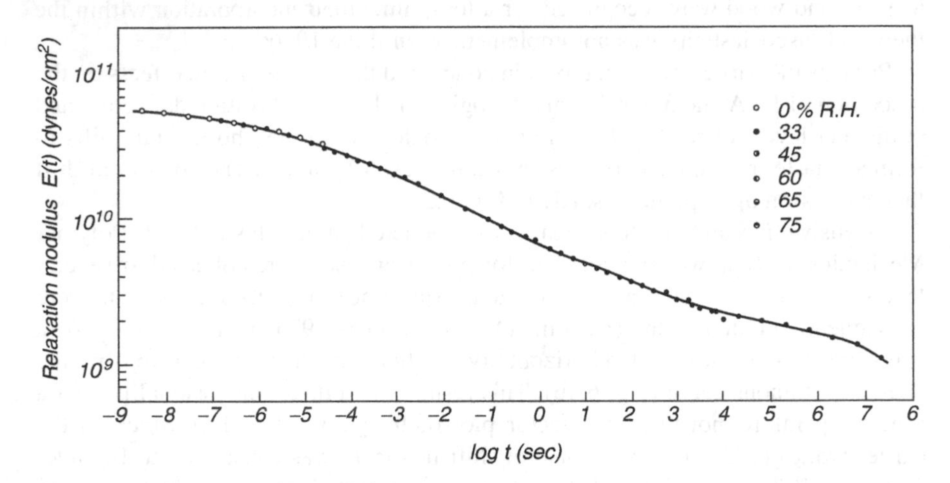

Fig. 6.2 A coalescent master relaxation modulus curve vs. log *t*
Copyright (Onogi S, Sasaguri K, Adachi T, and Ogihara S (1962). "Time-humidity superposition in some crystalline polymers." *Journal of Polymer Science*, 58(166):1–17) "This material is reproduced with permission of John Wiley & Sons, Inc."

With a_{Tm} at hand, it was possible to express the creep compliance $D(t) = D_o + \Delta D(t)$, where

$$\Delta D(t) = \sum a_i(1 - \exp(\xi/\tau_i)). \tag{6.2}$$

Note that the above representation of a_{Tm}, which is always larger than unity, differs from the more common convention where $\xi = t/a_{Tm}$, where $0 < a_{Tm} \leq 1$.

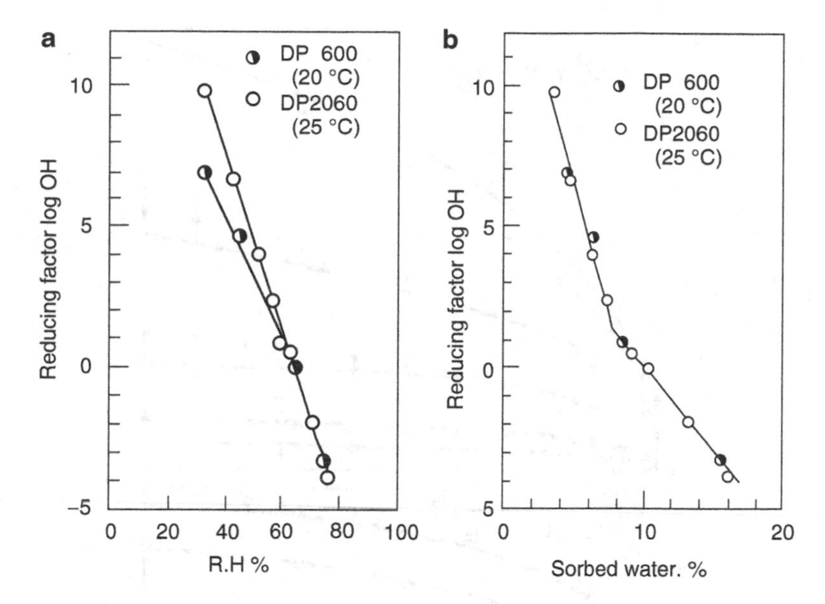

Fig. 6.3 Horizontal shift factors vs. relative humidity and absorbed water content for PVA nylon six films of two degrees of polymerization
Copyright (Onogi S, Sasaguri K, Adachi T, and Ogihara S (1962) "Time-humidity superposition in some crystalline polymers." *Journal of Polymer Science*, 58(166): 1–17) "This material is reproduced with permission of John Wiley & Sons, Inc."

The effects of moisture and temperature on creep under combined tension and shear should, in principle, reflect the presence of two distinct shift functions. Nevertheless, it may be argued (Maksimov et al. 1975, 1976a, b) that if the Poisson's ratio, ν, exhibits weak dependence on m and T, both the shear and Young's moduli G and E are nearly proportional to each other (since $E = 2(1 + \nu)G$) and could be associated with the same shift factor function a_{Tm}. For the polyester resin at hand, it was noted that ν varied between 0.44 and 0.48, thereby suggesting that, as a reasonable approximation, a_{Tm} expressed by (6.1) could be employed under combined loading as well.

The validity of the above approximation was tested by comparing long-term creep data under constant values of applied shear and tensile stresses, while fluctuating both moisture and temperature, with computational predictions. Results are shown in Fig. 6.5.

Note that at least part of the reasonably small discrepancy between data and model predictions in Fig. 6.5 must be attributed to the diffusion process delay time between fluctuations in ambient humidity and the attainment of uniformity in moisture distribution. This may explain why predictions tend to lag behind experimental results.

The same basic method was subsequently applied to characterize the response of fiber-reinforced polymeric composites. As expected, fluid effects on creep were pronounced in matrix-dominated orientations and negligible parallel to the fiber directions. Typical results, akin to those shown in Fig 6.4a, are exhibited in Fig. 6.6

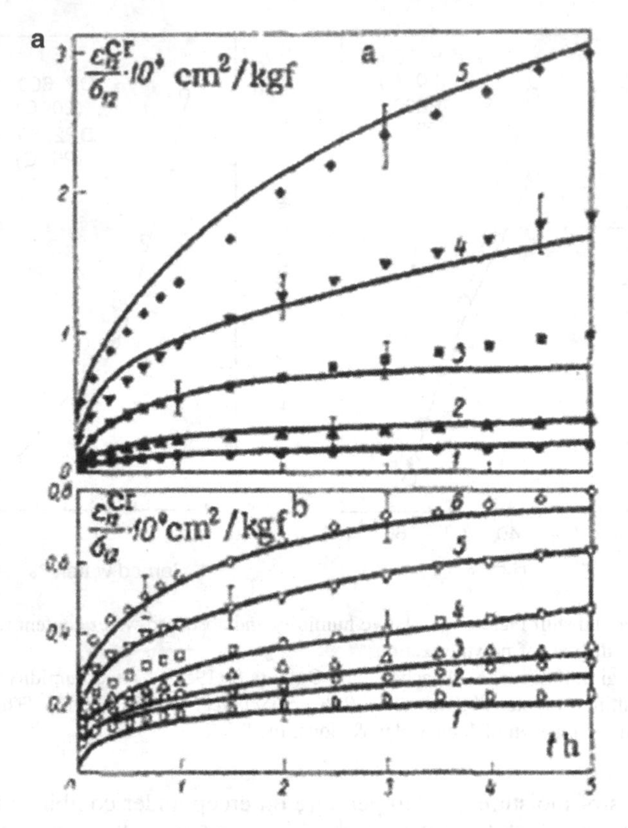

Fig. 6.4 (**a**) Compliance curves for shear creep: (a) $m = 0.2\%$, $T = 20$ (1), 30 (2), 40 (3), 50 (4) and 60°C (5); (b) $T = 20°\text{C}$, $m = 0.2$ (1), 0.5 (2), 0.7 (3), 1.0 (4), 1.5 (5), and 2.0% (6). Points, experiments; lines, calculations based on hygrothermal Viscoelastic model. (Maksimov et al. 1976a, b, Fig. 1). With kind permission from Springer Science+Business Media. (**b**) Compliance in shear creep (a) generalized compliance curved reduced to $T_o = 20°\text{C}$ and $m_o = 0.7\%$; the notation of the points correspond to Fig 6.4a (b) dependence of ln a on T for $m_o = 0.2\%$ (1) and on m for $T_o = 20°\text{C}$ (2); points, experiment; lines, calculations based on a non-product form of a shift factor expression. (Maksimov et al. 1976a, b, Fig. 3). With kind permission from Springer Science+Business Media: Mechanics of Composite Materials, "Influence of temperature and humidity on the creep of polymer materials. 3. Shear, and Shear and Tensile Strain Acting Together." 12(4), 1976, pp. 562-567, Maksimov et al. 1976a, b, Fig. 3

for a $[\pm 45]_{2S}$ lay-up of AS/3502 graphite/epoxy composite, where the logarithm of the compliance s_{xx} is plotted vs. log t for various levels of temperature and moisture content (Kibler 1980). These results could then be shifted horizontally to form master curves for s_{xx} and, upon employing laminate theory, for the transverse compliance s_T. These are shown in Fig. 6.7a, b. Inversely, results for the relaxation modulus of the same lay-up for T300/934 graphite epoxy are shown in Fig. 6.8 (Crossman et al. 1978), with "master relaxation curves" for three different graphite/epoxy composites of the same lay-up shown in Fig. 6.9. The master curves were

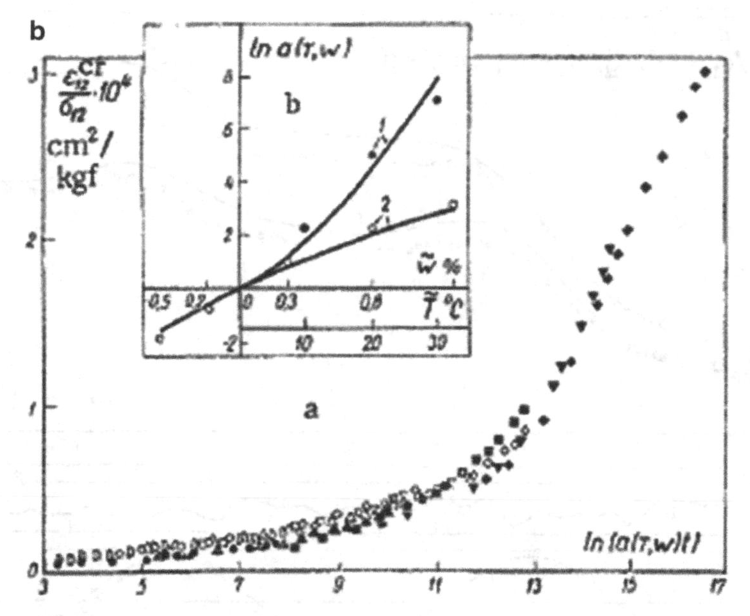

Fig. 6.4 (continued)

formed by employing the shift factors $a_{Tm}(T, m)$ exhibited in Figs. 6.10–6.12. Similar data for AS1/3501-6 are shown in Figs. 6.13 and 6.14 (Crossman and Warren 1985).

Employing the data for a_{Tm}, and representing the "master relaxation modulus" by means of a Prony series as shown in Fig. 6.15 (Crossman and Flaggs 1979) enabled to computational evaluation of the warping of a nonsymmetric $[0°_4/90°_4]_T$ laminate exposed to moisture and temperature (Crossman and Flaggs 1978; Flaggs and Crossman 1981). Results concerning the effects of fluctuating moisture and temperatures on the response of a polymeric slab were reported elsewhere (Weitsman 1977).

More recent publications in a similar vein (Patankar et al. 2008; Tsai et al. 2009) indicate that the issue of moisture effects on time-dependent material response retains its importance with the advent of new materials and novel applications.

6.2 Residual Hygrothermal Stresses. Data and Analysis

Composite materials are cured at elevated temperatures that may range between 175 and 190°C for epoxy-based materials and could exceed 450°C for polyimide resin components. Cooling down to room temperature gives rise to residual thermal stresses on both the laminate and individual ply levels. While the laminate-level residual stresses were discussed in Chap. 2, the ply-level stresses arise due to the disparity in thermal expansion coefficients between fiber and resin. The evaluation of those stresses is hampered by the paucity of data on the coefficient of the transverse expansion of fibers. Nevertheless, it was possible to demonstrate the

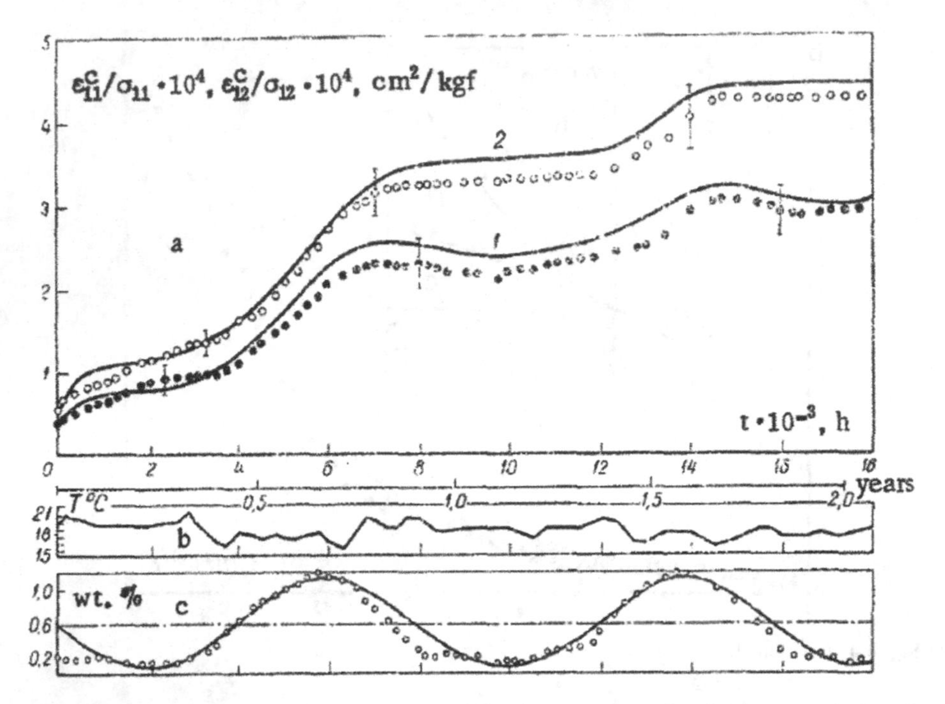

Fig. 6.5 Experimental (points) and calculated (lines) creep curves for simultaneous shear and tension under conditions of changing humidity of PN-3 resin [a: 1) ε^c_{11}; (2) ε^c_{12}], experimental values of temperature (**b**), and experimental values (points) and approximations (lines) of humidity of the material (**c**). (Maksimov et al. 1976a, b, Fig. 6). With kind permission from Springer Science +Business Media:

general features of those tresses, as sketched in Fig. 6.16a below (Crossman and Warren 1985). A similar sketch can be drawn under the ingress of moisture, upon its attaining a uniform, equilibrium distribution profile.

The above figures demonstrate that both sets of residual stresses tend to counteract each other, which is due to the fact that cool-down is associated with shrinkage, whereas the uptake of fluid causes expansion.

Nevertheless, a straight forward superposition of the profiles drawn in Figs. 6.16a, b would fail to represent realistic circumstances due to the huge disparity, of about six orders of magnitude, between the coefficients of thermal and moisture diffusion. In fact, the exceedingly slow process of moisture diffusion would leave major portions of the composite ply in the dry, or nearly dry, state for long times and the non-uniformity in moisture profiles would modify the distribution of residual stress on the ply level, as shown in Fig. 5.7.

Another aspect of residual stresses stems from their sign reversal upon drying caused by viscoelastic relaxation that was already alluded to in Chap. 2 and sketched in Fig. 2.3. For the case of fluid ingress alone, this is demonstrated by the following example (Weitsman 1979).

Consider a thin adhesive layer confined between rigid adherends exposed to moisture at $X = 0$. In view of the imposed constraints, moisture-induced swelling

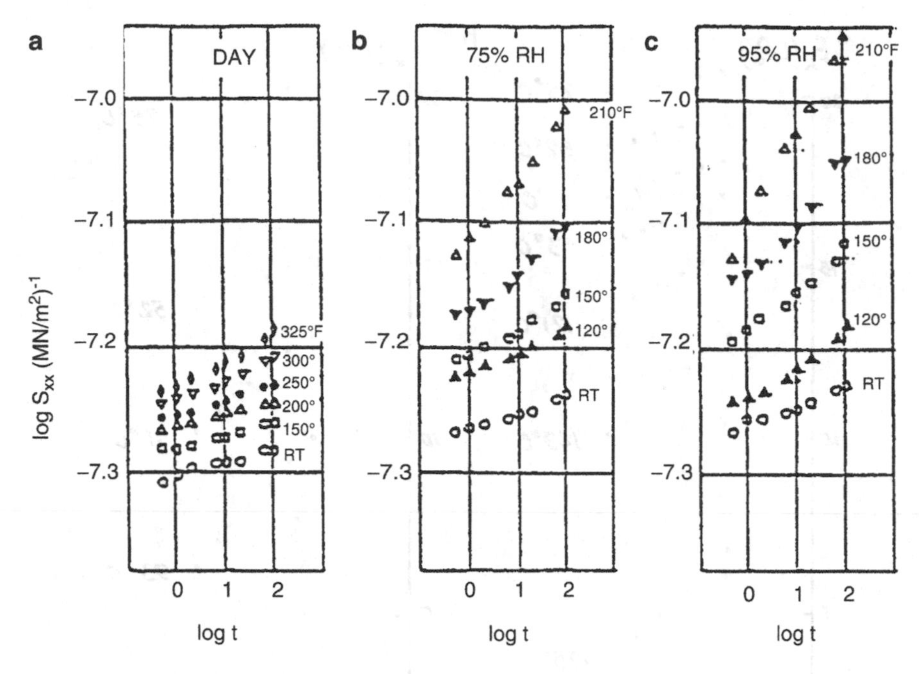

Fig. 6.6 The creep compliance S_{xx}, as recorded at various temperatures and relative humidities, for $[\pm 45]_{2s}$ AS/3502 composite specimens with (**a**) RH = 0%, (**b**) RH = 75% and (**c**) RH = 95%. (Kibler 1980)

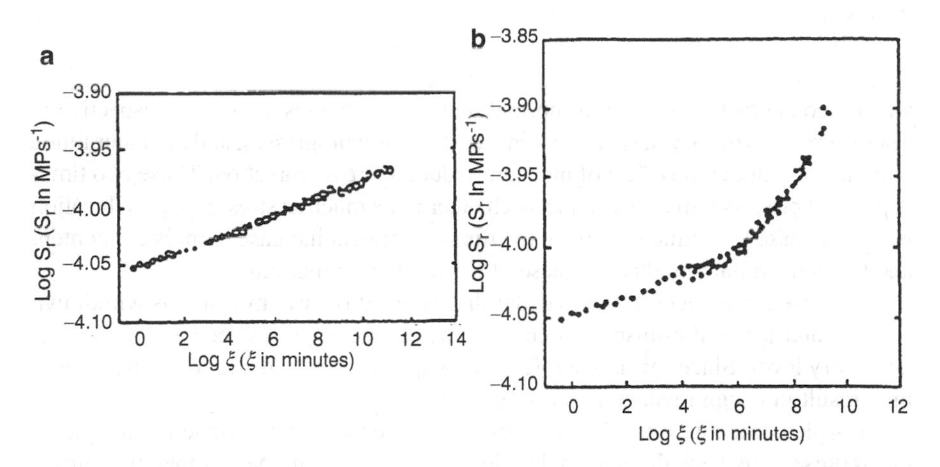

Fig. 6.7 "Master curves" for the transverse compliance S_T, as inferred from data for S_{xx} shown in Fig. 6.6 for AS/3502 composites. Curves obtained by horizontal shifts (parallel to the *log t* axis) of isothermal data. For (**a**) dry (**b**) "wet" cases. (Harper 1983)

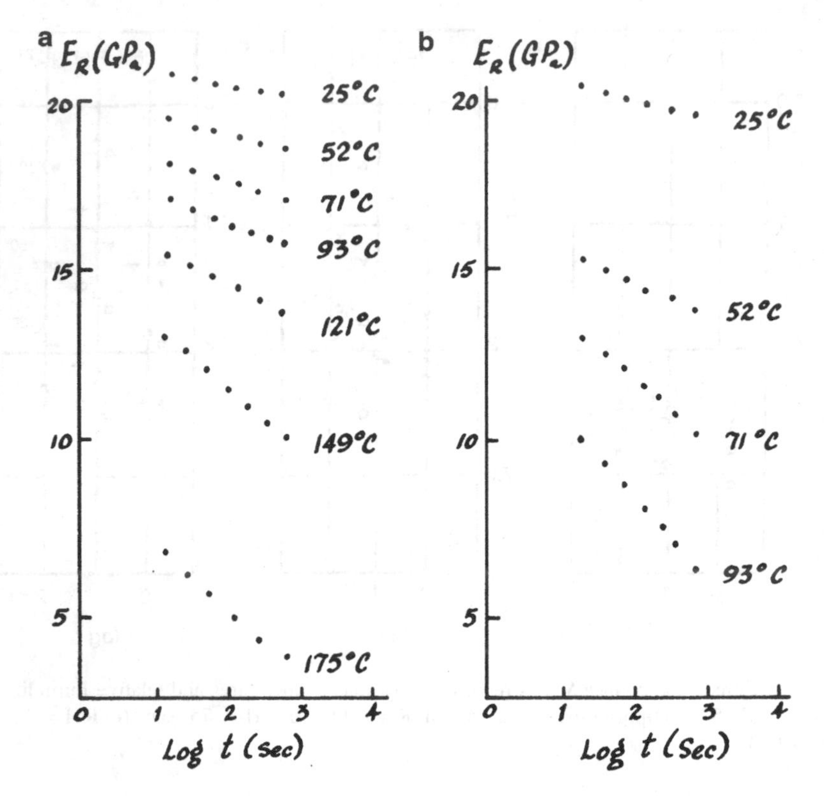

Fig. 6.8 Relaxation modulus of $[\pm 45]_{2s}$ T300/934 laminates containing (**a**) 0.14% moisture and (**b**) 1.40% moisture. (Crossman et al. 1978)
Credit Line: Reprinted, with permission, from "Moisture altered viscoelastic response of graphite/epoxy composite." copyright ASTM International, 100 Barr Harbor Drive, West Conshohocken, PA 19428

will be counteracted by internal normal and shear stresses s_n and s_t, respectively. These stresses will vary in space and time as diffusion progresses, and their amplitude will diminish under the effect of moisture-induced stress relaxation. Those two time-dependent processes may counteract each other in as much as stresses at each location would increase with time in view of the time-dependent increase in moisture content and the concomitant gradual decrease of the relaxation modulus.

A reverse phenomenon occurs under drying. In this case, moisture is withdrawn and, by and large, diminished in content, while the modulus recovers toward its larger dry level. Since the above effects are superimposed on a relaxed stress field, they result in a sign reversal in stresses.

This phenomenon, typical to viscoelastic response under reverse inputs, yields the representative results shown in Fig. 6.17. Note that the comparative linear elastic results tend toward a horizontal asymptopic value of approximately −0.075 during absorption and recover toward zero following desorption.

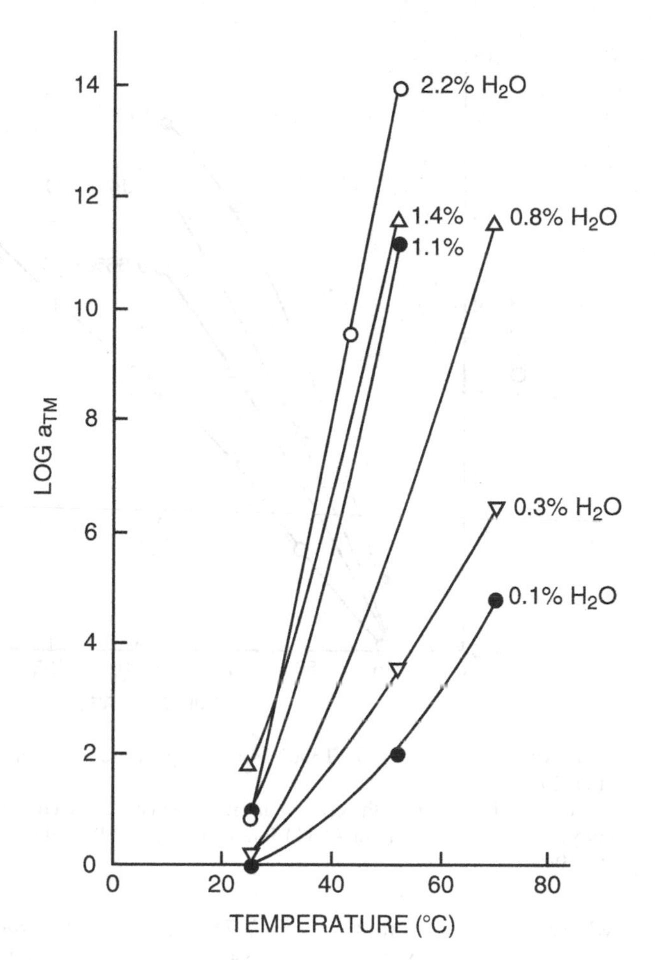

Fig. 6.9 Shift factor a_{TM} for GY70/339 vs. temperature and moisture. (Crossman et al. 1978)
Credit Line: Reprinted, with permission, from "Moisture altered viscoelastic response of graphite/epoxy composite." copyright ASTM International, 100 Barr Harbor Drive, West Conshohocken, PA 19428

A more detailed outline of hygrothermal viscoelastic analysis is provided below for the case of a 12-ply antisymmetric lay-up $[0/90/0_4/90_4/0/90]_T$ of AS4/3502 graphite/epoxy composite (Harper 1983; Harper and Weitsman 1985).

The master curves shown in Figs. 6.4a, b were generated by means of the shift factor $a_{TH}(T,H) = a_T(T) \cdot a_H(H)$, where $a_T = \exp(B - T/A)$ and $a_4 = \exp(c_0 m + c_1)$ with $B = 45.81, A = 6.258, c_0 = 5.2, c_1 = 0.26$, T in °K, and m in % weight gain.

All of the viscoelasticity was considered to dwell in the transverse compliance[1] S_T which, upon curve fitting of creep data could be expressed by

$$S_T = \begin{cases} S(t + t_0)^q & \text{dry state,} \\ S_0 + S_1(t/A_{TH})^n & \text{wet state.} \end{cases} \tag{6.3}$$

[1] The compliances S_L, S_T, and S_{LT} are the reciprocals of Q_{11}, Q_{22}, and Q_{12} in Chap. 2, respectively.

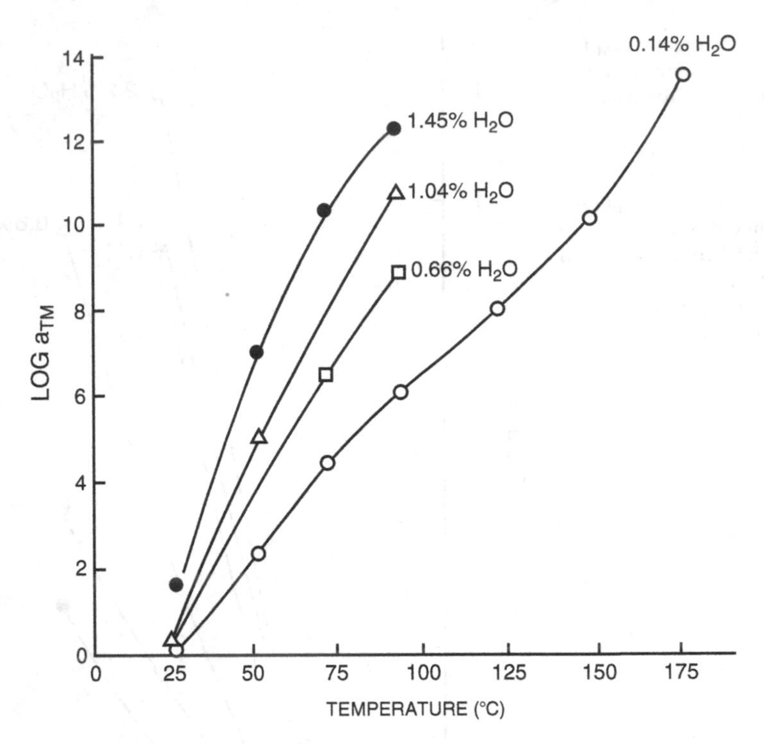

Fig. 6.10 Shift factor a_{TM} for T300/934 vs. temperature at several moisture contents. (Crossman et al. 1978)
Credit Line: Reprinted, with permission, from "Moisture altered viscoelastic response of graphite/epoxy composite." copyright ASTM International, 100 Barr Harbor Drive, West Conshohocken, PA 19428

where $t_0 = 1$ min, $q = 7.75 \times 10^{-3}$, $S = 88.2 \times 10^{-6}\ \text{MPa}^{-1}\ \text{min}^{-q}$, $S_0 = 88.2 \times 10^{-6}\ \text{MPa}^{-1}$, $n = 0.14$, $S_1 = 1.662 \times 10^{-6}\ \text{MPa}^{-1}\ \text{min}^{-n}$ and time t is in minutes.

The time-independent longitudinal and transverse compliance were $S_L = 8 \times 10^{-6}\ \text{MPa}^{-1}$ and $S_{LT} = 2.32 \times 10^{-6}\ \text{MPa}^{-1}$.

Experiments were performed with several replicate square plates of dimensions 10 × 10 cm (4″ × 4″) and 6.25 × 6.25 cm (2.5″ × 2.5″). Those were exposed to a controlled regimen of varying temperature and relative humidity, with weight gain data collected periodically. In view of the aforementioned antisymmetry of the lay-up, all plates deformed into anti-clastic shapes with initial curvatures k_i caused by thermal cool down. These initial curvatures were predicted quite well by linear elastic laminate theory.[2]

[2] It should be noted that plates of larger dimensions exhibited geometric instability. They tended to deform into cylindrical shapes that could be snapped through between these two principal directions.

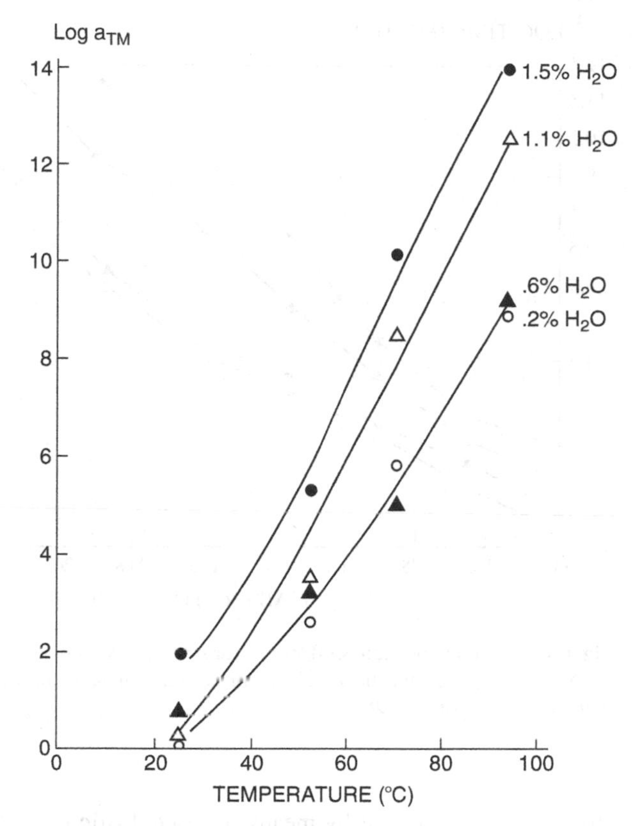

Fig. 6.11 Shift factor a_{TM} for T300/5209 vs. temperature at several moisture contents. (Crossman et al. 1978) Credit Line: Reprinted, with permission, from "Moisture altered viscoelastic response of graphite/epoxy composite." copyright ASTM International, 100 Barr Harbor Drive, West Conshohocken, PA 19428

Upon subsequent exposure to ambient environments in several humidity and temperature-controlled chambers these curvatures varied with time due to the two aforementioned contradictory time-dependent mechanisms. While the time-dependent moisture diffusion process serves as a stress-inducing mechanism, the time-dependent relaxation acts to reduce the level of those stresses. Note that relaxation depends on moisture content through the shift factor $a_H(m)$.

In calculating the time-dependent curvatures, it was first necessary to evaluate the spatial and temporal variations of moisture profiles in relation to the controlled fluctuation in ambient humidity. Moisture levels $m(z, t)$, where z denotes coordinate across the laminate's thickness, were then incorporated within the shift factor $a_H(m)$ at sufficiently small time intervals[3] at each level of the coordinate z. Subsequently, it was possible to compute time variation of the in-plane stresses at all z corresponding to the distinct histories of the moisture content therein. These

[3] It was found advantageous to utilize equal intervals of $\log(t)$.

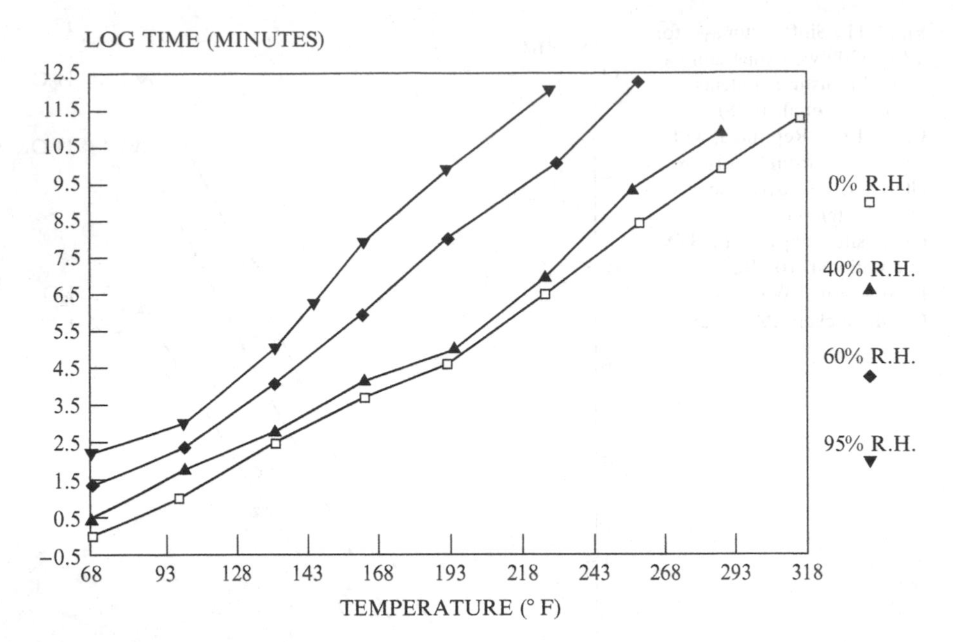

Fig. 6.12 Time-temperature shift factors for stress relaxation of AS1/3501-6 as a function of prior exposure to equilibrium moisture content at constant temperature/relative humidity conditions (Crossman and Warren 1985)

stresses were evaluated by means of viscoelastic convolution integrals that were incorporated within laminate theory. The latter step yields the values of curvature as well (Harper 1983).

Typical stress profiles across the thickness of the laminate at several instances into the cyclic exposure regime are shown in Fig. 6.18a–g, where linear elastic values are shown by the dashed lines. Note that the viscoelastic stresses exceed their elastic counterparts at various locations and times.

Predicted elastic and viscoelastic curvatures are depicted in Fig. 6.19. These are plotted for the nondimensional departure of the time-dependent curvature $k(t)$ from its initial value k_i, namely $h[k_i - k(t)]$ vs. time t. Note the overshoot in the predicted viscoelastic values.

Both elastic and viscoelastic predictions depart noticeably from experimental measurements, especially during desorption. This disparity is mostly due to the onset of profuse microcracks within the laminate, such as that shown in Fig. 4.30a–c. These fractures reduce the overall level of the energy stored within the laminate and bring about an increase in curvature (i.e., a trend toward a flatter laminate shape).

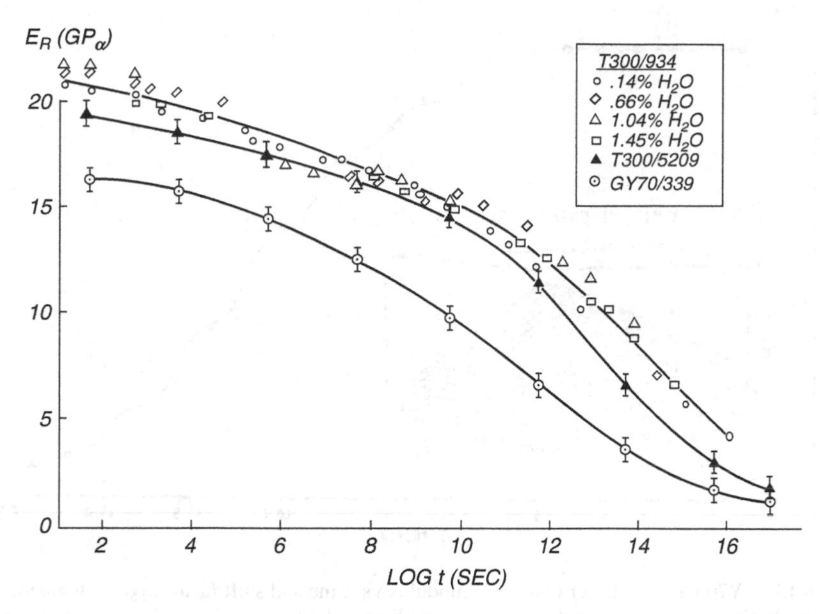

Fig. 6.13 Master relaxation modulus curves for $[\pm 45]_{2s}$ laminates of three composite materials. Horizontally shifted T300/934 data are shown. (Crossman et al. 1978)
Credit Line. Reprinted, with permission, from "Moisture altered viscoelastic response of graphite/epoxy composite." copyright ASTM International, 100 Barr Harbor Drive, West Conshohocken, PA 19428

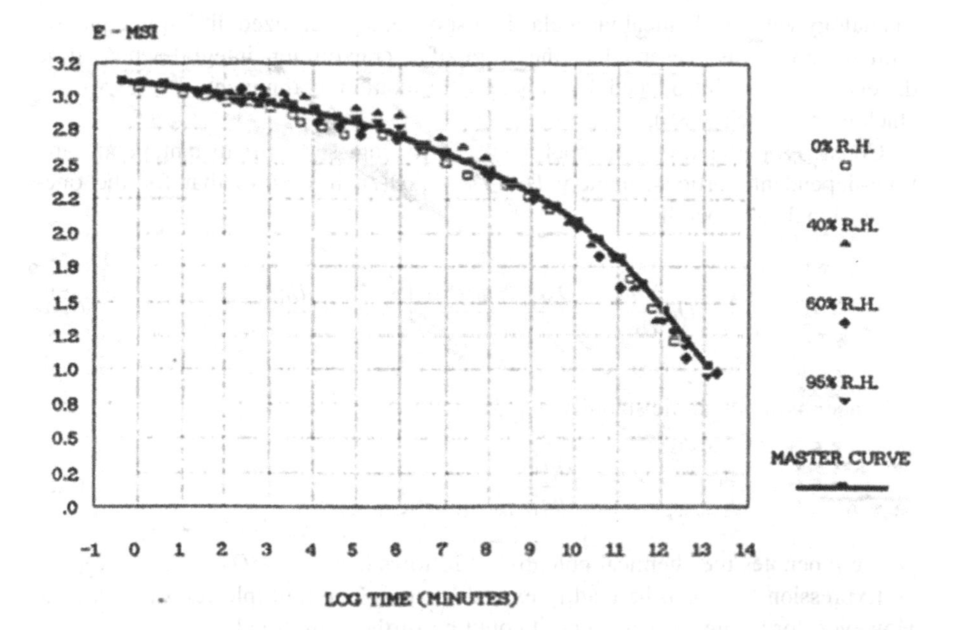

Fig. 6.14 The master time-humidity stress relaxation curve for AS1/3501-6. Tensile stress relaxation modulus vs. Log time. (Crossman and Warren 1985)

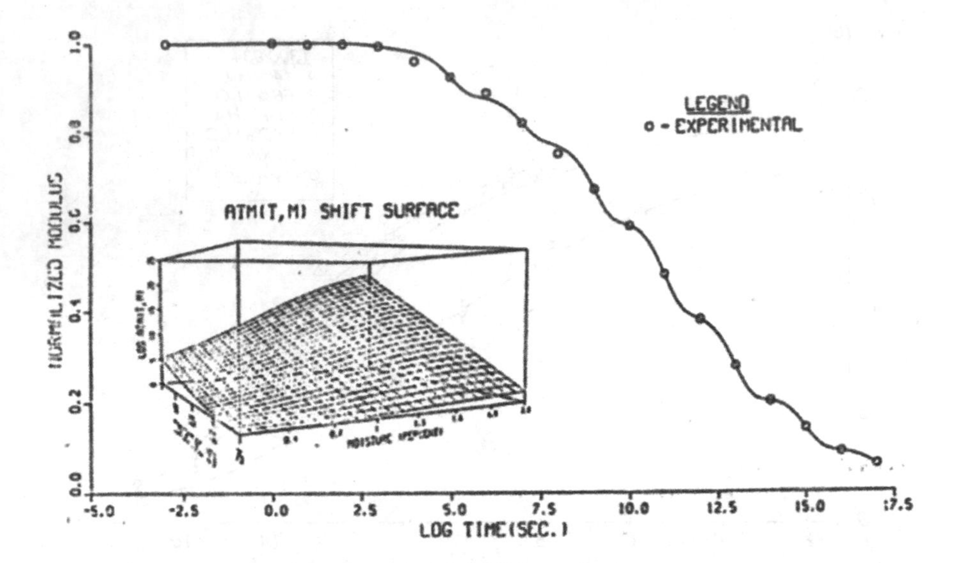

Fig. 6.15 GY70/CE339 Master relaxation modulus vs. time and shift factor a_{TM} vs. temperature and equilibrium moisture content. (Crossman and Flaggs 1979)
Credit: Crossman F. W. Flaggs D.L. (1979). "Dimensional Stability of Composite Laminates During Environmental Exposure," SAMPE Journal, Vol. 15, 15–20. Reprinted by permission of SAMPE

6.3 Viscoelastic Modeling

In analogy with mechanical viscoelastic response, a generalized diffusion equation with memory was presented in the form of a convolution integral where time dependence was introduced by a single relaxation component $\phi = \exp(-\gamma\tau)$ (Jackle and Frisch 1986).

Considering the diffusion coefficient D to consist of an instantaneous and time-dependent portions, namely $D = D_0 + \Delta D(t)$, it follows that for the one-dimensional case, one has

$$\frac{\partial m}{\partial t} = D_0 \frac{\partial^2 m}{\partial x^2} + \frac{(D_\infty - D_0)\partial^2}{\partial x^2} \int_{-\infty}^{t} \phi(t-s) \frac{\partial m(x,s)}{\partial s} \mathrm{d}s. \tag{6.4}$$

Since thermodynamic stability requires that

$$\left.\frac{\partial \mu}{\partial m}\right|_{t\to\infty} \geq \left.\frac{\partial \mu}{\partial m}\right|_{t=0} \tag{6.5}$$

where μ denotes the chemical potential, it follows that $D_\infty \geq D_0$.

Expression (6.4) can be readily extended to include multiple relaxation times. However, for a single component, it could be further simplified.

An analogous expression was presented for the chemical potential μ.

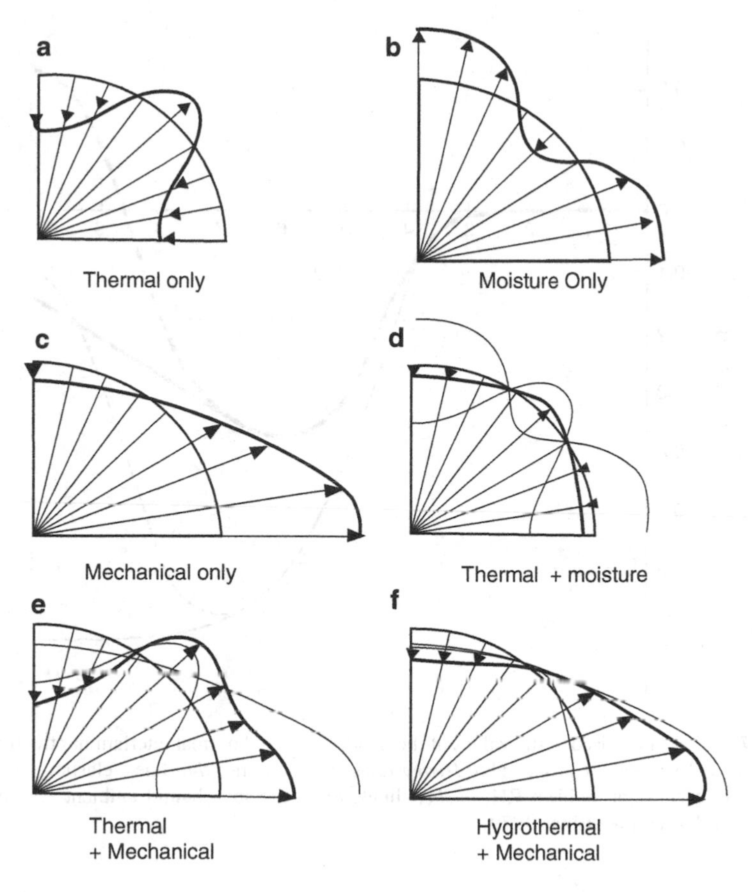

Fig. 6.16 Schematic view of fiber-matrix interface normal stress state with the length and direction of vectors indicating the relative magnitude and sign of the stress at selected positions on the interface. (Crossman and Warren 1985)

A different approach (Weitsman 1990) aimed at the establishment of a viscoelastic diffusion model on the basis of fundamental principles of irreversible thermodynamics in combination with the formalization of continuum mechanics. This approach is akin to that presented in Sect. 5.6 and 5.7.

Starting again with (5.15) through (5.20), it is possible to represent that latter inequality by means of the Gibbs free energy $\phi = u - TS - \sigma_{ij} \cdot \varepsilon_{ij}$ as follows:

$$-\rho_{s0}\phi - \rho_{s0}s\dot{T} - \dot{\sigma}_{ij}\varepsilon_{ij} - (q_i/T)g_i + \tilde{\mu}\dot{m} - f_i\tilde{\mu}_{,i} - \tilde{s}g_if_i \geq 0. \tag{6.6}$$

Following the previously developed schemes of Biot (1954) and Schapery (1964, 1966, 1969), while employing standard index notation that leads to a systematic employment of material symmetries, adding the nondimensional

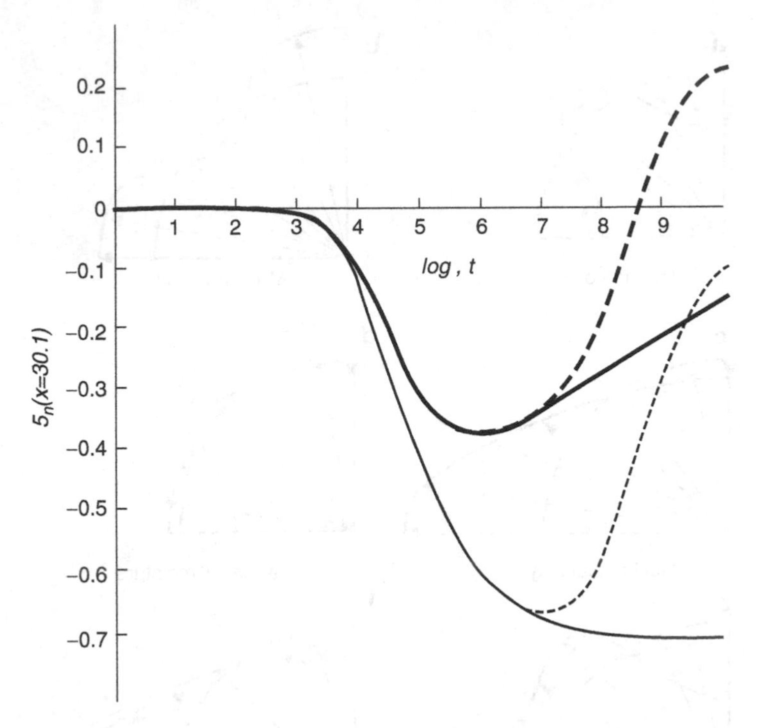

Fig. 6.17 Elastic and Viscoelastic values of the nondimensional normal interlaminar traction S_n at $x = (X/a) = 3$ vs. log t (t in seconds). *Heavy* lines, Viscoelastic; *thin* lines, elastic; *solid* lines, exposure to a constant ambient RH; *dashed* lines, exposure to a humid ambient environment followed by drying. (Weitsman 1979)

fraction of free volume v_f as an internal variable that accounts for aging and incorporating fluid content m, it is considered that

$$\phi = \phi(\sigma_{ij}, m, T, \gamma_r, v_f). \tag{6.7}$$

In (6.7), γ_r ($r = 1,2,\ldots,N$) are scalar internal variables that account for the internal configurations of the polymeric material.

Since (6.6) cannot be violated by any process, the methodology of continuum mechanics yields the following constitutive equations:

$$\varepsilon_{ij} = -\rho_{s0}\frac{\partial\phi}{\partial\sigma_{ij}}, \tag{6.7.1}$$

$$s = -\frac{\partial\phi}{\partial T}, \tag{6.7.2}$$

$$\mu = \rho_{s0}\frac{\partial\phi}{\partial m} \tag{6.7.3}$$

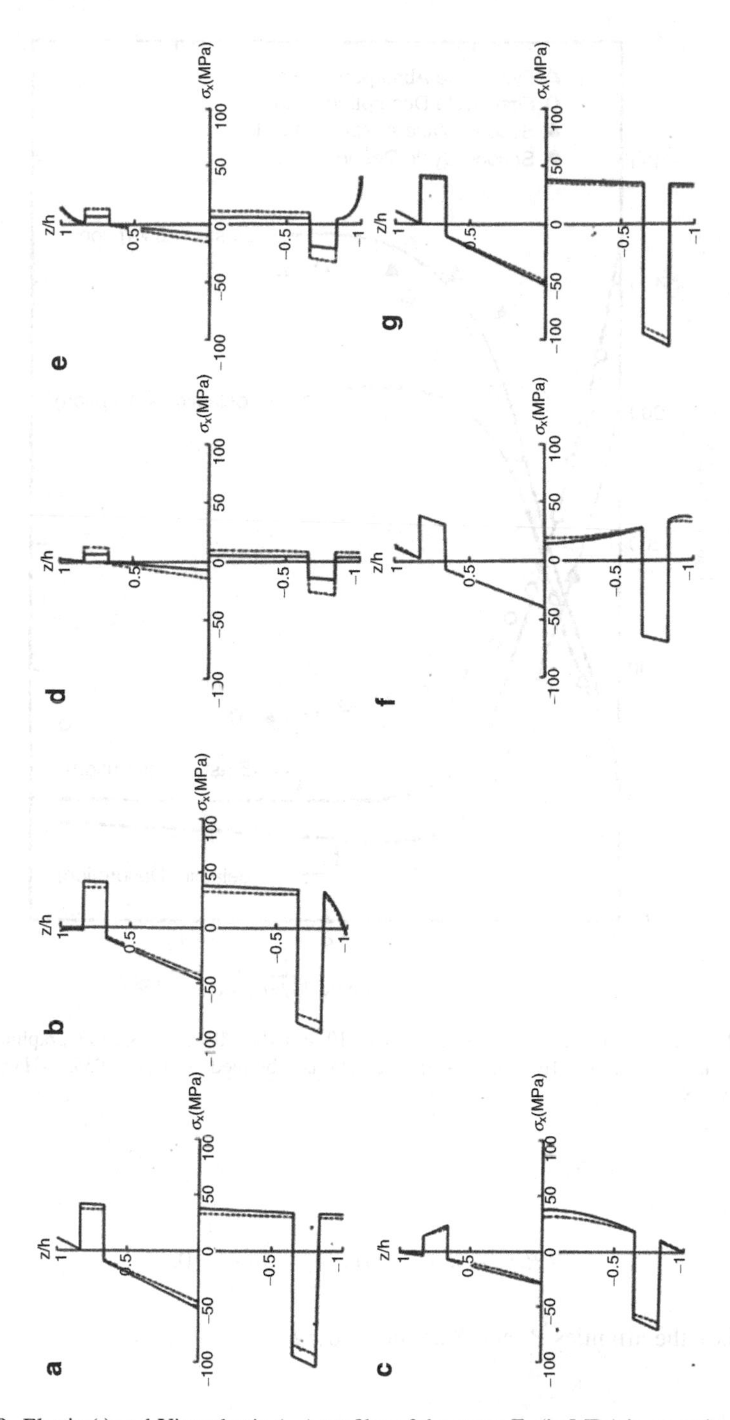

Fig. 6.18 Elastic (·) and Viscoelastic (—) profiles of the stress T_g (in MPa) in an antisymmetric $[0/90/0_4/90_4/0/90]_T$ AS/3502 composite laminate at various stages of exposure to ambient relative humidity. (**a**) After a 4h linear cool down from the cure temperature of 452 K (355°F) to the conditioning temperature of 339 K (150°F). (b) After 0.323 days of conditioning at RH = 95%. (**c**) After 9.76 days at RH = 95%. (**d**) At saturation. (**e**) After 0.081 days of drying at RH = 0% past saturation. (**f**) After 16.96 days of drying at 0% RH past saturation. (**g**) At total desorption past saturation. (Harper 1983)

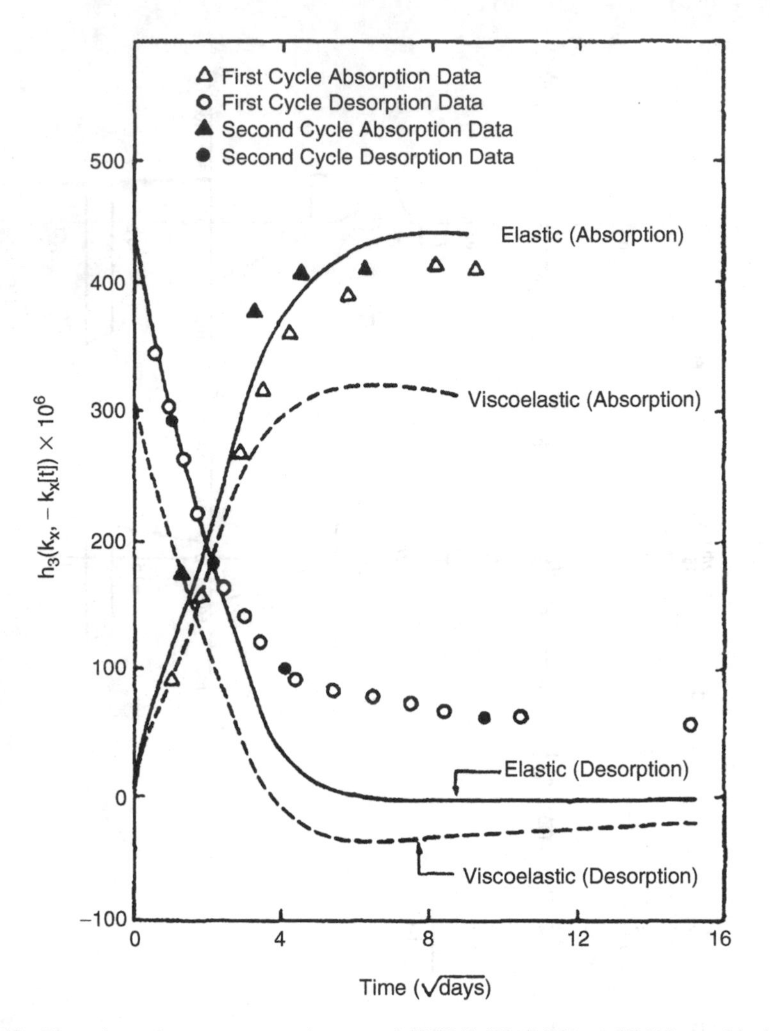

Fig. 6.19 Time-dependent curvature changes of $[0/90/0_4/90_4/0/90]_T$ AS/3502 graphite/epoxy laminates at $T = 346$ K (163°F). Absorption data is obtained at RH = 95%. (Harper and Weitsman 1985)

and

$$-R_r\dot{\gamma}_\gamma - R\dot{v}_\mathrm{f} - (q_i/T)g_i - \tilde{s}g_i f_i \geq 0. \tag{6.8}$$

In (6.8), the affinities R_r and R are defined by

$$R_r = \rho_{s0}\frac{\partial\phi}{\partial\gamma_r}; \quad R = \rho_{s0}\frac{\partial\phi}{\partial v_\mathrm{f}}. \tag{6.9}$$

As noted in Chap. 2, the aging process is associated with the spontaneous collapse of the free volume v_f from an initial value v_{f0} which is created by the cooling of a polymer across its glass transition temperature T_g down to a temperature T_0. In addition, the subjection of a polymer externally to σ_{ij}, m or $\Delta T = T - T_0$ will trigger an irreversible process, which will cause the internal variables γ_r and v_f to drift spontaneously toward their equilibrium values γ_r^e and v_f^e. The two above-mentioned processes are assumed to be independent of each other (Struik 1978; Ferry 1980). Consequently, while both $\dot{\gamma}_v$ and $\dot{v}_f$ in (6.8) denote derivatives with respect to time, the time lapse t for γ_r is not necessarily equal to the time span τ for v_f.

Assume that $\gamma_r, v_f \ll 1$. A Taylor expansion about the initial state ($\gamma_r = 0, v_f = v_{f0}$) then reads

$$\rho_{s0}\phi = \rho_{s0}\left[\phi_0 + \phi_r\gamma_r + \phi_v(v_f - v_{f0}) + \frac{1}{2}\phi_{rq}\gamma_r\gamma_q + \frac{1}{2}\phi_{vv}(v_f - v_{f0})^2 + \text{H.O.T.}\right]. \tag{6.10}$$

In (6.10), $\phi_0 = \phi_0(\sigma_{ij}, m, T, 0, v_{f0})$ and $\phi_r, \phi_v, \phi_{rq}, \phi_{vv}$ denote partial derivatives at $\gamma_r = 0, v_f = v_{f0}$, all of which depend on $\sigma_{ij}, m,$ and$T$.

According to the extremum principles of thermodynamics (Callen 1960) ϕ attains its minimum at the equilibrium values $\gamma_r = \gamma_r^e$ and $v_f = v_f^e$. Therefore, at equilibrium we have

$$\frac{\partial\phi}{\partial\gamma_r} = 0, \quad \frac{\partial\phi}{\partial v_f} = 0,$$

$$\frac{1}{2}\frac{\partial^2\phi}{\partial\gamma_r\partial\gamma_s}\delta\gamma_r\delta\gamma_s + \frac{1}{2}\frac{\partial^2\phi}{\partial v_f^2}(\delta v_f)^2 > 0$$

for all $\delta\gamma_r$ and δv_f. Expanding ϕ about γ_r^e and $v_f^e = 0$ we have

$$\rho_{s0}\phi = \rho_{s0}\left[\phi_e + \frac{1}{2}\phi_{rq}^e(\gamma_\gamma - \gamma_\gamma^e)(\gamma_q - \gamma_q^e) + \frac{1}{2}\phi_{vv}(v_f^e)^2 + \text{H.O.T.}\right]. \tag{6.11}$$

In view of the assumption that $\gamma_r, \gamma_r^e, v_f, v_{f0} \ll 1$, expansions (6.10) and (6.11) are equal, implying in particular that

$$\phi_{vv} = \phi_{vv}^e, \quad \phi_{rq} = \phi_{rq}^e.$$

Therefore, $\phi_{vv} > 0$ and ϕ_{rq} are components of a symmetric, positive definite matrix.

Employing the familiar assumption of viscous-like resistance let

$$R = -b(m, T, v_f)\frac{dv_f}{d\tau} \tag{6.12}$$

and

$$R = -b_{rq}(m, T, v_f)\frac{d\gamma_q}{dt}. \quad (6.13)$$

By Onsager's principle $b_{rq} = b_{qr}$ and in view of the dissipation inequality (6.8), $b>0$ and b_{rq} are components of a semipositive definite, symmetric matrix. As noted earlier, the times τ and t in (6.12) and (6.13) are distinct because the quench time τ usually occurs earlier than the exposure time t, namely $t = \tau - t_e$.

Equations (6.9), (6.11) and, (6.12) give

$$\rho_{s0}\phi^{e}_{vv}v_f + b\frac{dv_f}{d\tau} = 0, \quad (6.14)$$

where by

$$v_f = v_{f0}e^{-\zeta/\tau_v} \quad (6.15)$$

with $d\zeta = d\tau/b$ and $\tau_v^{-1} = \rho_{s0}\phi^{e}_{vv}$. Equations (6.9), (6.10) and, (6.13) yield

$$\rho_{s0}\phi_{rq}\gamma_q + b_{rq}\dot{\gamma}_q = -\rho_{s0}\phi_r. \quad (6.16)$$

Since ϕ_{rq} and b_{rq} are symmetric, positive definite matrices (the case of a semipositive definite b_{rq} is omitted here for the sake of brevity; this case corresponds to a viscoelastic fluid, as modeled by a free dash pot attached to Kelvin or Maxwell elements), they can be diagonalized simultaneously and (6.16) is expressible in the form

$$\rho_{s0}\Phi_p\hat{\gamma}_p + B_p\dot{\hat{\gamma}}_p = -\rho_{s0}\phi_p \quad (p = 1, \ldots, N;\ \text{no sum on } p). \quad (6.17)$$

Note that b_{rq}, and therefore B_p, depend on the quench time τ as well as on the exposure time t through their dependence on m,T, and v_f in (6.13).

Following Schapery assume that all ϕ_{rq} have a *common* dependence on $\sigma_{ij}, m,$ and$T$, namely $\phi_{rq} = a_G(\sigma_{ij}, m, T){\phi_{rq}}^0$ with constant ${\phi_{rq}}^0$ and, similarly $b_{rq} = a(m, T, v_f){b_{rq}}^0$. These common "shift factors" a_G and a carry over to (6.17) whose solution, subject to the conditions $\gamma_r(0) = 0, \gamma_r(\infty) = \gamma_r^e$, reads

$$\hat{\gamma}_p = -K_p(1 - e^{-\xi/\tau_v}), \quad (6.18)$$

where

$$K_p = \frac{\phi_p}{a_G\Phi_p}, \quad \tau_p = \frac{B_p}{\rho_{s0}\Phi_p} \quad (\text{no sum on } p, p = 1, 2, \ldots, N) \quad (6.19)$$

and

$$\mathrm{d}\xi = \left(\frac{a_G}{a}\right) \mathrm{d}t. \tag{6.20}$$

Aiming at linear stress-strain behavior, consider $\rho_{s0}/E \ll 1$, where E denotes a characteristic modulus. To retain linearity assume that all retardation times do not depend on σ_{ij}. Hence, in (6.15) and (6.18), $\tau_v = (m, T, v_f)$ and $\tau_p = (m, T, v_f), p = 1, \ldots, N$. Furthermore, to establish linear stress-strain relations, consider the expansion in (6.10) truncated after three terms, which, resorting to the diagonalized form, reads

$$\rho_{s0}\phi = \rho_{s0}[\phi_0 + \phi_r\hat{\gamma}_r + \phi_v(v_f - v_{f0})]. \tag{6.21}$$

Equation (6.7) gives

$$\varepsilon_{ij} = -\rho_{s0}\left[\frac{\partial\phi_0}{\partial\sigma_{ij}} + \frac{\partial\phi_r}{\partial\sigma_{ij}}\hat{\gamma}_r + \frac{\partial\phi_v}{\partial\sigma_{ij}}(v_f - v_{f0})\right]. \tag{6.22}$$

The specific forms that follow from (6.22) depend on the material symmetry at hand. Considering isotropy we have

$$\phi(\sigma_{ij}, m, T, \hat{\gamma}_r, v_f) = \phi(\sigma_{kk}, \sigma_{ij}\sigma_{ij}, m, T, \hat{\gamma}_r, v_f)$$

(the stress invariant $|\sigma_{ij}|$ is inadmissible for linear behavior) which, for linear behavior, gives the well-known forms

$$\begin{aligned}
\rho_{s0}\phi_0 &= A_0(m,T) - L_0(m,T)\sigma_{kk} - M_0(m,T)\sigma_{kk}\sigma_{ll} - N_0(m,T)\sigma_{kl}\sigma_{lk},\\
\rho_{s0}\phi_r &= A_r(m,T) - L_r(m,T)\sigma_{kk} - M_r(m,T)\sigma_{kk}\sigma_{ll} - N_r(m,T)\sigma_{kl}\sigma_{lk},\\
\rho_{s0}\phi_v &= A_v(m,T) - L_v(m,T)\sigma_{kk} - M_v(m,T)\sigma_{kk}\sigma_{ll} - N_v(m,T)\sigma_{kl}\sigma_{lk}.
\end{aligned} \tag{6.23}$$

It is advantageous to express L_0, L_r, and L_v in terms of expansional coefficients α and β, namely

$$L(m,T) = \alpha(m,T)\Delta T + \beta(m,T)m. \tag{6.24}$$

Equations (6.22) and (6.23) then yield

$$\begin{aligned}
\varepsilon_{ij} = {} & [\hat{\alpha}_0 - \hat{\alpha}_r\hat{\gamma}_r - \hat{\alpha}_v(v_f - v_{f0})]\Delta T\delta_{ij} + [\hat{\beta}_0 - \hat{\beta}_r\hat{\gamma}_r - \hat{\beta}_v(v_f - v_{f0})]m\delta_{ij} + 2\\
& \times [\hat{M}_0 - \hat{M}_r\hat{\gamma}_r - \hat{M}_v(v_f - v_{f0})]\sigma_{kk}\delta_{ij} + 2[\hat{N}_0 - \hat{N}_r\hat{\gamma}_r - \hat{N}_v(v_f\\
& - v_{f0})]\sigma_{ij}.
\end{aligned} \tag{6.25}$$

It can be noted from (6.18) and (6.19) that depends on σ_{ij} through the presence of ϕ_p in K_p. However, for sufficiently small stresses, K_p themselves can be expanded according to (6.23), whereby upon collecting like powers in σ_{ij}, the *form* of the strain-stress relationship given in (6.25) remains valid with stress-independent $\hat{\gamma}_r$.

The stress-strain relations (6.25) are of the familiar form employed in linear viscoelasticity, except that the retardation spectra incorporate now aging affects, and all instantaneous compliances age with time. Recalling (6.18) for a discrete spectrum of retardation times τ_r, expressions (6.25) read

$$\begin{aligned}\varepsilon_{ij} = &\left\{\left[\alpha_0 - \alpha_1(1 - e^{-\zeta/\tau_v}) + \sum_r \alpha_r(1 - e^{-\xi/\tau_r})\right]\Delta T\right.\\ &+ \left[\beta_0 - \beta_1(1 - e^{-\zeta/\tau_v}) + \sum_r \beta_r(1 - e^{-\xi/\tau_r})\right] m\\ &\left.+ \left[B_0 - B_1(1 - e^{-\zeta/\tau_v}) + \sum_r B_r(1 - e^{-\xi/\tau_r})\right]\sigma_{kk}\right\}\delta_{ij}\\ &+ \left[J_0 - J_1(1 - e^{-\zeta/\tau_v}) + \sum_r J_r(1 - e^{-\xi/\tau_r})\right]\sigma_{ij}. \qquad (6.26)\end{aligned}$$

In (6.26), the quantities $\alpha, \beta, B,$ and J depend on m and T, while the "reduced times" ξ and ζ both depend of m, T, and the ageing time τ. However, in contrast to ζ, the reduced time ξ is measured from the time of application of $\sigma_{ij}, m,$ and$T$. For fluctuating environments and stresses, expression (6.26) takes the form of convolution integrals. The case of a continuous spectrum of retardation times τ_r can be represented in the familiar manner by means of retardation integrals.

Consider now stress-assisted diffusion in aging viscoelastic media. Employing (6.7), (6.21), (6.23), and (6.24) we obtain

$$\begin{aligned}\mu = &\frac{\partial A_0}{\partial m} + \frac{\partial A_r(m,T)}{\partial m}\gamma_r + \frac{\partial A_v(m,T)}{\partial m}(v_f - v_{f0})\\ &+ \left\{\left[-\frac{\partial \alpha_0}{\partial m} + \frac{\partial \alpha_r}{\partial m}\gamma_r \frac{+\partial \alpha_v}{\partial m}(v_f - v_{f0})\right]\Delta T - \frac{\partial(m\beta_0)}{\partial m} + \frac{\partial(m\beta_r)}{\partial m}\gamma_r \frac{+\partial(m\beta_v)}{\partial m}(v_f - v_{f0})\right\}\sigma_{kk}\\ &+ \left[-\frac{\partial M_0}{\partial m} + \frac{\partial M_r}{\partial m}\gamma_r \frac{+\partial M_v}{\partial m}(v_f - v_{f0})\right]\sigma_{kk}\sigma_{ll} + \left[-\frac{\partial N_0}{\partial m} + \frac{\partial N_r}{\partial m}\gamma_r \frac{+\partial N_v}{\partial m}(v_f - v_{f0})\right]\sigma_{kl}\sigma_{lk}.\end{aligned} \qquad (6.27)$$

The boundary conditions for moisture transport are

$$\mu(x,t) = \mu_A(\boldsymbol{x},t) \quad (\mathbf{x}\text{ on boundary}), \qquad (6.28)$$

where μ_A represents the chemical potential of the ambient vapor.

$$
\begin{aligned}
\frac{\partial A_0}{\partial m} = \mu_A &+ \sum_r W_r(m,T)(1-e^{-\xi/\tau_r}) + W(m,T)(1-e^{-\zeta/\tau_v}) \\
&+ \left[-\beta_0 + \sum_r \beta_r(m,T)(1-e^{-\xi/\tau_r}) + \beta_v(m,T)(1-e^{-\zeta/\tau_v})\right]\sigma_{kk} \\
&+ \left[-Y(m,t) + \sum_r Y_r(m,T)(1-e^{-\xi/\tau_r}) + Y_v(m,T)(1-e^{-\zeta/\tau_v})\right]\sigma_{kk}\sigma_{ll} \\
&+ \left[-Z(m,t) + \sum_r Z_r(m,T)(1-e^{-\xi/\tau_r}) + Z_v(m,T)(1-e^{-\zeta/\tau_v})\right]\sigma_{kl}\sigma_{lk}
\end{aligned}
\tag{6.29}
$$

Assume for simplicity that the expansional coefficients α and β do not vary with m. In this case, (6.27) and (6.28) together with (6.18) yield

Equation (6.29) illuminates the effects of viscoelastic retardation and aging, as well as stress, on the diffusion process. In the absence of those effects, the boundary condition, which reads $\partial A_0/\partial m = \mu_A$, translates into the familiar statement $m(x,t) = m_0(t)$, $\mathbf{x}$ on boundary and m_0 prescribed. Therefore, the term μ_A corresponds to a classical process of diffusion through a "mechanically inert" solid. On the other hand, (6.29) states that when the equilibrium boundary value is approached gradually with time (even for exposure to constant ambient vapor pressure), it is affected by age and depends quadratically on the applied stress.

The second term on the right side of (6.29) expresses the effects of viscoelastic, time-dependent motions of the polymeric network on the moisture absorption process, while the third term represents the role of free volume on diffusion.

Consider now the diffusion coefficient D. Since this coefficient reflects microlevel phenomena, its form cannot be derived from a continuum model. (The principle of equipresence suggests that $D = D(\sigma_{ij}, m, T, \gamma_r, v_f)$ but provides no further insight.) Molecular-level considerations of diffusion in solids and fluids suggest that

$$D \sim \text{vacancy in host material} \times \text{agitation energy of diffusing substance}$$

In the present case, it is reasonable to expect that the vacancy would depend mostly on v_f and secondarily on m, T, and σ_{ij} (through their influence on swelling strains.) On the other hand the agitation energy would depend mostly on T.

Accordingly

$$D \sim D_0\left[m, T, v_f(\xi(\tau)), \int_0^t F(\xi(\tau) - \xi(\eta))\frac{d\sigma_{kk}}{d\eta}d\eta\right] \times (\exp(-U/RT)). \tag{6.30}$$

In the absence of a satisfactory molecular theory for the glassy state at the present time, it appears expedient to base the form of D on empirical evidence.

In isotropic materials, the flux of moisture at fixed σ_{ij} and T is taken to be (when m, T and σ_{ij} vary in space, couplings occur between the fluxes of moisture and temperature and both depend on the gradients of μ, T, and the invariants of stress).

$$f_i = -D\frac{\partial \mu}{\partial x_i}. \tag{6.31}$$

Consequently, balance of mass yields the field equation

$$\frac{\partial m}{\partial t} = \frac{\partial}{\partial x_i}\left(D\frac{\partial \mu}{\partial x_i}\right). \tag{6.32}$$

In view of (6.27) and (6.30), relations (6.31) and (6.32) are extremely complicated even at $T = T_0$ and $\sigma_{ij} = \sigma_{ij}{}^0$; the complexity is due to the presence of m in D and v_f (through $\xi(\tau)$ and $\zeta(\tau)$ in the compliances listed in (6.27). However, in spite of the cumbersome details and the paucity of information, several features of the form of (6.32) emerge. Accordingly, the diffusion equation contains (1) nonlinear terms in the moisture gradient $\partial m/\partial x_i$; (2) follows a time-retardation process akin to mechanical viscoelastic response; (3) varies nonlinearly with external stresses σ_{ij}; (4) exhibits an aging behavior characteristic of glassy polymers.

It is interesting to note that the case of two-phase diffusion can be expressed in the context of the thermodynamic theory developed in this work. In this case, let

$$\rho_{s0}\phi = \rho_{s0}\phi(\sigma_{ij}, m_\mathrm{m}, m_\mathrm{b}, T, \gamma_r, v_\mathrm{f}) \tag{6.33}$$

Or, since $m_\mathrm{m} + m_\mathrm{b} = m$

$$\rho_{s0}\phi = \rho_{s0}\phi(\sigma_{ij}, m, m_\mathrm{m}, T, \gamma_r, v_\mathrm{f}). \tag{6.33a}$$

In analogy with (6.7), we now have

$$\mu = \rho_{s0}\frac{\partial \phi}{\partial m}, \quad \mu_\mathrm{m} = \rho_{s0}\frac{\partial \phi}{\partial m_\mathrm{m}}.$$

Equation (6.27) and the boundary condition (6.28) pertain now to the chemical potential of the mobile phase μ, so that all partial derivatives in those equations should be taken with respect to m_m as well instead of m alone.

Similarly, the flux-gradient relation now involves the mobile phase only, so that (6.31) is replaced by

$$f_i = -D\frac{\partial \mu_\mathrm{m}}{\partial x_i}. \tag{6.34}$$

Since the mass balance relation remains valid, (6.22) is replaced now by

$$\frac{\partial m}{\partial t} = \frac{\partial}{\partial x_i}\left(D\frac{\partial \mu_\mathrm{m}}{\partial x_i}\right). \tag{6.35}$$

The system of equations is supplemented by the phase interaction relation (3-28b) as before.

The dependence of field equations and boundary conditions on retardation times, physical ageing, and stress is retained for the case of two-phase diffusion and remains analogous to (6.27)–(6.29).

It is worth noting that the predicted dependence of μ on the first and second power of stress, expressed in (6.27), is reflected in the data shown in Fig. 4.21.

While expressions of a continuum viscoelastic damage relations are now available (Abdel-Tawab and Weitsman 1998; Smith and Weitsman 1999), these did not incorporate the presence of fluids. While such a development may follow the approach presented herein, no such formulation is available as of now.

6.4 Nonlinear Characterization

It was noted that the time-temperature-moisture characterization of a polyimide film during stress relaxation varied with the level of applied strain, implying a nonlinear effect. Furthermore, while plots of the time-dependent portion $\Delta E(t)$ of the relaxation modulus $E(t) = E_e + \Delta E(t)$ collected at various fixed level of RH could be coalesced by horizontal shifts alone to form a master curve, this could not be achieved for $E(t)$ as a whole. It was therefore necessary to incorporate a vertical shift into the formulation and write

$$E(t, T, m, \varepsilon) = g(T, m, \varepsilon) E^R \left(\frac{t}{a(T, m, \varepsilon)} \right), \tag{6.36}$$

where g accounts for the vertical shift that was noted to reside in E_e alone, and E^R denotes the reference value at $T = T_R, m = 0$ and $\varepsilon \ll 1$, i.e., with $g = 1$ and $a = 1$.

Note that the diffusion process within thin films approaches equilibrium in very short times, thereby enabling the correlation between E_e and m, including the dependence of g on m. For the specific material at hand, additional experimental evidence suggested that E_e and g depended on m alone, thus the effect of moisture could be further isolated.

It followed that for the material at hand, the shift factor function a depended on T and ε alone, which could then be evaluated independently of m.

Altogether, expression (6.36) could be further reduced to

$$E(t, T, m, \varepsilon) = E_e(m) + g(m) \Delta E^R \left(\frac{t}{a(T, \varepsilon)} \right). \tag{6.37}$$

The specific expressions for $E_e(m)$, $\Delta E^R(t), g(m)$, and $a(T, \varepsilon)$ were obtained from data fitting procedure and are listed in that work (Harper et al. 1997).

References

Abdel-Tawab K, Weitsman YJ (1998) A Coupled Viscoelasticity/Damage Model with Application to Swirl-Mat Composites. International Journal of Damage Mechanics 7(4):351–380
Biot MA (1954) Theory of Stress-Strain Relations in Anisotropic Viscoelasticity and Relaxation Phenomena. Journal of Applied Physics 25(11):1385
Callen HB (1960) Thermodynamics. Wiley, New York
Crossman FW, Flaggs DL (1978) Viscoelastic analysis of hygrothermal altered laminate stresses and dimensions, Final report funded in part by Air Force Office of Scientific Research under contract F49620-77-C-0122 and in part by Lockhead Independent Research Program
Crossman FW, Flaggs DL (1979) Dimensional Stability of Composite Laminates During Environmental Exposure. SAMPE Journal 15:15–20
Crossman FW, Warren WJ (1985) The influence of environment on matrix dominated composite fracture. Final report N60921-81-C-0157, Dec 1985
Crossman FW, Mauri RE, Warren WJ (1978) Moisture altered viscoelastic response of graphite/epoxy composite. In: Vinson JR (ed) *Advanced Composite Materials - Environmental Effects*, ASTM STP 658. American Society for Testing and Materials, Philadelpia, PA, pp 205–220
Ferry JD (1980) Viscoelastic Properties of Polymers. Wiley, New York
Flaggs DL, Crossman FW (1981) Analysis of the Viscoelastic Response of Composite Laminates During Hygrothermal Exposure. Journal of Composite Materials 15(1):21–40
Harper BD (1983) On the Effects of Post Cure Cool Down and Environmental Conditioning on Residual Stresses in Composite Laminates. Texas A & M University Report MM-4665-83-11, Aug 1983
Harper BD, Weitsman Y (1985) On the effects of environmental conditioning on residual stresses in composite laminates. International Journal of Solids and Structures 21(8):907–926
Harper B, Rao J, Kenner V, Popelar C (1997) Hygrothermal effects upon stress relaxation in a polyimide film. Journal of Electronic Materials 26(7):798–804
Jackle J, Frisch HL (1986) Properties of a generalized diffusion equation with a memory. J Chem Phys 85(3):1621–1627
Kibler KG (1980) Time-Dependent Environmental Behavior of Graphite/Epoxy composites. *Final Report*, General Dynamics Corporation, Fort Worth, TX, Contract F33615-77-C-5 109, Report AFWAL-TR-80-4052, February 1980
Maksimov RD, Mochalov VP, Urzhumtsev YS (1972) Time — Moisture superposition. Mechanics of Composite Materials 8(5):685–689
Maksimov RD, Sokolov EA, Mochalov VP (1975) Effect of temperature and humidity on the creep of polymer materials. Uniaxial elongation under variable temperatur e - humidity conditions. Mechanics of Composite Materials 11(6):834–839
Maksimov RD, Mochalov VP, Sokolov EA (1976a) Influence of temperature and humidity on the creep of polymer materials. 3. Shear, and Shear and Tensile Strain Acting Together. Mechanics of Composite Materials 12(4):562–567
Maksimov RD, Mochalov VP, Sokolov EA (1976b) Influence of temperature and humidity on the creep of polymeric materials. 4. Prediction on the basis of field test results. Mechanics of Composite Materials 12(6):859–864
Onogi S, Sasaguri K, Adachi T, Ogihara S (1962) Time-humidity superposition in some crystalline polymers. Journal of Polymer Science 58(166):1–17
Patankar K, Dillard D, Case S, Ellis M, Lai Y, Budinski M, Gittleman C (2008) Hygrothermal characterization of the viscoelastic properties of Gore-Select® 57 proton exchange membrane. Mechanics of Time-Dependent Materials 12(3):221–236
Schapery RA (1964) Application of Thermodynamics to Thermomechanical, Fracture, and Birefringent Phenomena in Viscoelastic Media. Journal of Applied Physics 35(5):1451
Schapery RA (1966) An engineering theory of nonlinear viscoelasticity with applications. International Journal of Solids and Structures 2(3):407–425

Schapery RA (1969) On the characterization of nonlinear viscoelastic materials. Polymer Engineering & Science 9(4):295–310

Smith L, Weitsman Y (1999) The visco-damage mechanical response of swirl-mat composites. International Journal of Fracture 97(1):301–319

Struik LCE (1978) Physical Aging in Amorphous Polymers and Other Materials. Elsevier, New York

Tsai Y, Bosze E, Barjasteh E, Nutt S (2009) Influence of hygrothermal environment on thermal and mechanical properties of carbon fiber/fiberglass hybrid composites. Composites Science and Technology 69(3–4):432–437

Weitsman YJ (1977) Effects of Fluctuating Moisture and Temperature on the Mechanical Response of Resin-Plates. Journal of Applied Mechanics ASME 44(4):571–576

Weitsman YJ (1979) Interfacial stresses in viscoelastic adhesive-layers due to moisture sorption. International Journal of Solids and Structures 15(9):701–713

Weitsman YJ (1990) A continuum diffusion model for viscoelastic materials. J Phys Chem 94 (2):961–968

Chapter 7
Effects of Fluids on Mechanical Properties and Performance

7.1 Damage, Strength, and Durability

It was already noted that fluids introduce residual stresses into polymeric composites and affect mechanical fields by enhancing the creep and relaxation processes. In addition, it was remarked that fluids may degrade polymers and fiber–matrix interfaces by hydrolysis and chemical attack, induce osmotic pressure within interphase regions that contribute to the weakening of fiber matrix bond, and chemically decompose the fibers themselves, especially glass fiber. Thus, while most fluid-induced degradations are affected through the polymeric phase of the composite, as well as the interface/interphase domains, some degradation may be due to fluid effects on glass fibers. While the outcome of the former is revealed through the lowering of shear-driven properties, that of the latter is detected by reduction in tensile resistance. Tensile properties may be completely or partially irreversible (Ishak et al. 2000). Reductions of 30–50% in interfacial strength were reported for some graphite/epoxy systems (Kaelble et al. 1975).

The above effects are reflected in significant reductions in the strength of such polymers as epoxy and in the transverse and shear strengths of unidirectionally reinforced polymeric composites (Pomies et al. 1995; Hertz 1973; Kaminski 1973, Hofer et al. 1974, 1975; Verette 1975; Browning et al. 1976; Husman 1976; Dewimille et al. 1980; Shen and Springer 1981; Hamada et al. 1991; Juska 1993).

Typically, one observes degradation of between 25 and 80% in the strength of epoxies and reduction between 50 and 80% in the transverse strength of unidirectionally reinforced fibrous composites attributable to the effects of solvents. The above range in the reduced values of strengths corresponds to increasing levels of temperature. The reader should be aware that some of the data reported in the literature were complied under excessively high temperatures, and while such results may serve to illuminate fluid effects they may be caused primarily by temperature, ranging over levels that are well above those that are relevant to most engineering applications. A decrease of strength with increasing exposure

Y.J. Weitsman, *Fluid Effects in Polymers and Polymeric Composites*,
Mechanical Engineering Series, DOI 10.1007/978-1-4614-1059-1_7,

time, ranging between 20 and 50%, was noted for glass/polyester composites immersed in water at 30°C (Norwood and Marchant 1981).

Reductions of 10–30% in shear strength and between 25 and 50% in failure shear strains were noted for ±45% glass epoxy coupons immersed in sea water and in distilled water, with more substantial reductions associated with weight gains along curve "C" in Fig. 4.2 (Davies et al. 1998).

A wealth of data on the shear strength, debond length, the transverse tensile, and flexural strength of several composites in the dry state, and under immersion in distilled water as well as sea water at several temperatures are shown in Figs. 7.1 and 7.2 (Ramirez et al. 2008).

Lesser reductions were observed in quasi-isotropic laminates (Kriz and Stinchcomb 1982) and no diminutions occurred in the longitudinal properties of carbon fiber composites. Smaller reductions in transverse strength were also noted in several graphite and glass fiber polymeric composites immersed in sea water (Grant 1991).

More substantial delamination zones were observed under monotonically increased loads of quasi-isotropic carbon/epoxy laminate with moisture content $m = 0.7\%$ than in the dry case (Ogi et al. 1999) while a comparative drop of 5% was noted in the tensile strength of "wet" uniaxially reinforced Kevlar-graphite/epoxy hybrid composites (Haque et al. 1991). Similar conclusions are implied elsewhere (Ogi and Takao 1998). An indirect confirmation of that case is implied in the above article by the observation that when plotted vs. a monotonically increased stress, the Poisson's ratio grows faster in the wet case than its dry counterpart.

It is illuminating to remark that in some circumstances, strength reductions in E-glass/polyester sheet molding compounds (SMCR50) correlated with moisture weight loss that accords with curve "D" in Fig. 4.2 (Springer et al. 1981).

Though no information is available about the detailed nature of degradation mechanisms in SMC-R5O caused by water, it is interesting to note that distinct degradation mechanisms were observed in E-glass/polyester SMC composites exposed to chemical solutions. Acids induced stress corrosion cracking within the fibers (Somiya and Morishita 1993a), while alkalines degraded the fiber-resin interface (Somiya and Morishita 1993b). The latter circumstance resulted in a more severe reduction in the fracture toughness of the composite.

While carbon fibers are largely immune to solvents, glass fibers are highly susceptible to chemical attack by water, acids, and alkaline solutions (Thomas 1960; Metcalfe et al. 1971; Metcalfe and Schmitz 1972; Aveston et al. 1980; Bailey et al. 1980; Dewimille et al. 1980; Friedrich 1981; Aveston and Sillwood 1982; Jones et al. 1982; Hsu and Chou 1985; Sheard and Jones 1986; French and Pritchard 1991). A rational, though descriptive, explanation for the moisture-enhanced failure mechanisms in E-glass fibers was forwarded by McKinnis on the basis of experimental observations and basic considerations (McKinnis 1978). Accordingly, the failure of E-glass fibers is affected by the level of relative humidity of the ambient moisture as well as by the sizes and content of the impurities that are imparted to the glass during its manufacture. This susceptibility is noted by the formation of pits on

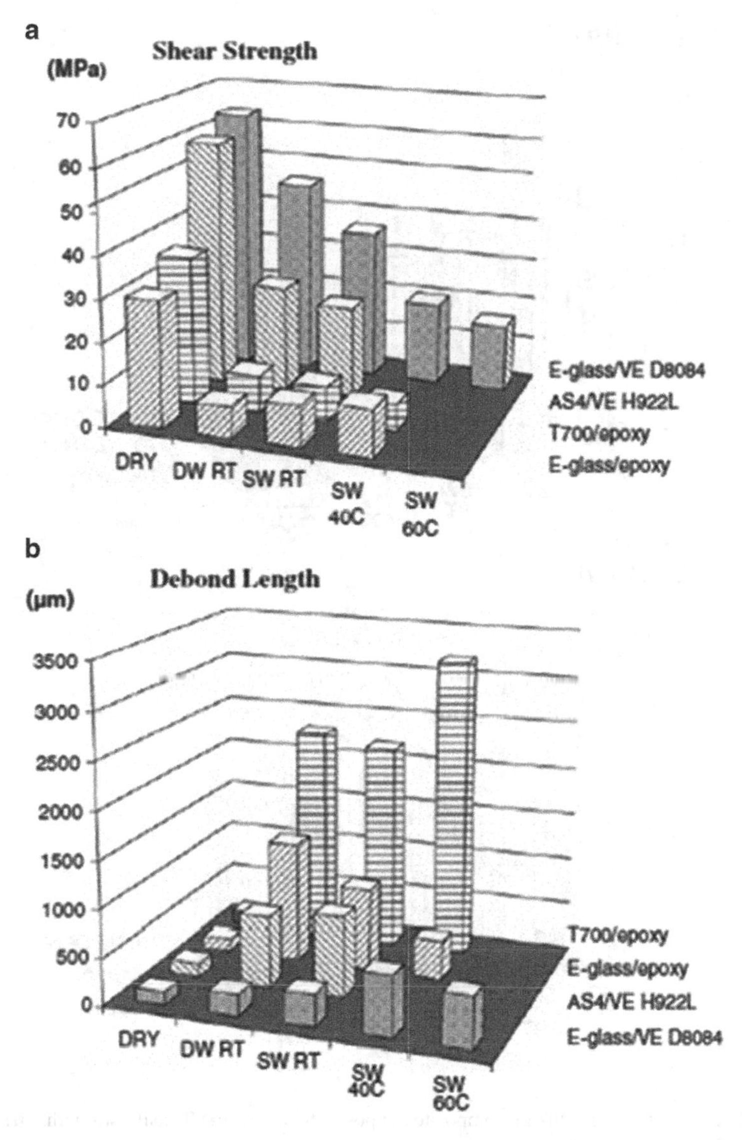

Fig. 7.1 F/M interface shear strength (**a**) and debond length (**b**) of various systems exposed to water (Ramirez et al. 2008)

the surface of glass fibers immersed in water (Ashbee and Wyatt 1969) and in sea water (White and Phillips 1985) and by the progression of stress-corrosion cracking, leading to significant reductions in durability, as reflected by shorter times-to-failure in static fatigue under exposure to water and acids (Charles 1958a, b; Aveston et al. 1980, 1982; Bailey et al. 1980; Dewimille et al. 1980; Hogg et al. 1981; Kelly and McCartney 1981; Hogg and Hull 1982; Aveston and Sillwood 1982; Jones et al. 1982; Hsu and Chou 1985; Sheard and Jones 1986; Castaing et al.

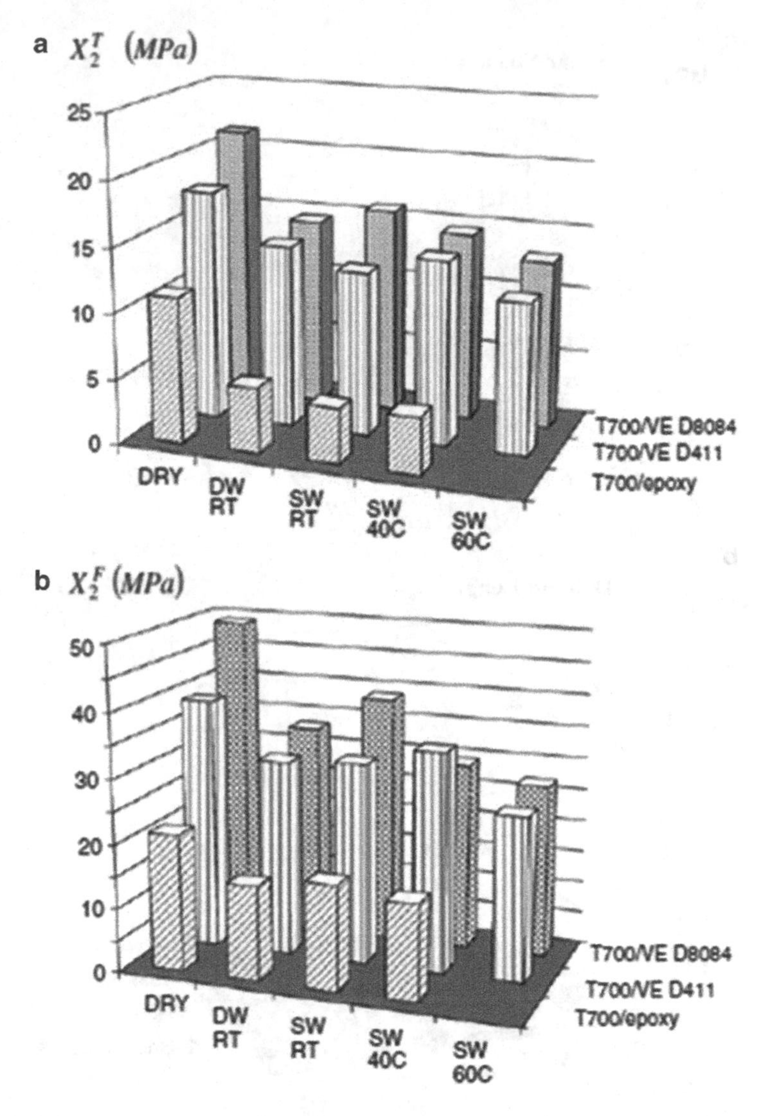

Fig. 7.2 Transverse strengths of composites exposed to water. (**a**) Tensile strength, (**b**) flexure strength (Ramirez et al. 2008)

1993; Chateauminois et al. 1993; Corum et al. 1998). As remarked earlier, the corrosion of glass fibers is delineated by a sharp front that separates the yet intact core from an outer, corroded, concentric cylindrical region. Reductions in the strength of glass fibers and their composites can be correlated with the advance of the above corrosion process (Ehrenstein and Spaude 1984).

It is worth noting that while data on times-to-failure under static fatigue are usually correlated by the expression (Zhurkov 1965)

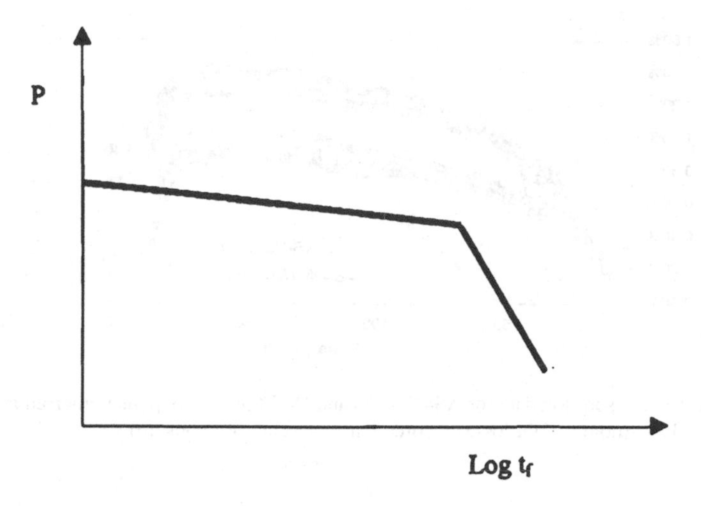

Fig. 7.3 A semilogarithmic plot of load P vs. failure time t_f in static fatigue tests of glass/polyester composites in sea water (adapted from White and Phillips 1985)

$$\sigma = A - B \log t_f, \tag{7.1}$$

plots of stress σ vs. failure time t_f in some composites, such as polyester/glass immersed in sea water, suggest the existence of two-stage mechanisms (White and Phillips 1985) as depicted in Fig. 7.3. This circumstance resembles the incubation times required for the stepping-up in creep response (Jain et al. 1979; Jain and Asthana 1980; Menges and Gitschner 1980; Bird and Allan 1981). It may also be noted that data on the static fatigue of glass-reinforced plastics immersed in sea water (Wyatt et al. 1981) seem to suggest the existence of a two-stage mechanism, although they were fitted by the authors according to (7.1). Note that the long-term weight gain data for immersed gr/ep composites, shown in Fig. 7.4, suggest the presence of a two-stage mechanism, perhaps the progression of a slow chemical reaction.

The foregoing two-stage mechanisms in the evolution of failure under static fatigue were observed also in R glass/epoxy specimens exposed to distilled water (Chateauminois et al. 1993). These mechanisms were interpreted in the context of the fatigue model (Talreja 1981, 1990) sketched in Fig. 7.5. Accordingly, the shorter lives at elevated stresses are due to the incidence of random fiber failures attributed to the statistical distribution of their initial flaw population. The subsequent gradual increase in times to-failure under lower levels of external loads is deemed to reflect the progressive failure within the composite as broken fibers transfer their loads to intact neighboring fibers. This interpretation may differ from the incubation- time suggestion.

It stands to reason that in the presence of fluids, both ε_e and ε_m will drop below their dry values.

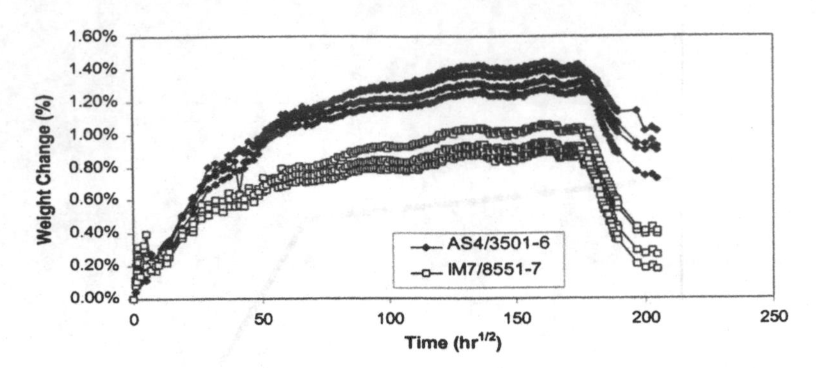

Fig. 7.4 Five year sorption data for AS4/3501-6 and IM7/8551-7 coupons immersed in simulated seawater at 34°C (note weight losses recorded after 4 years of exposure)

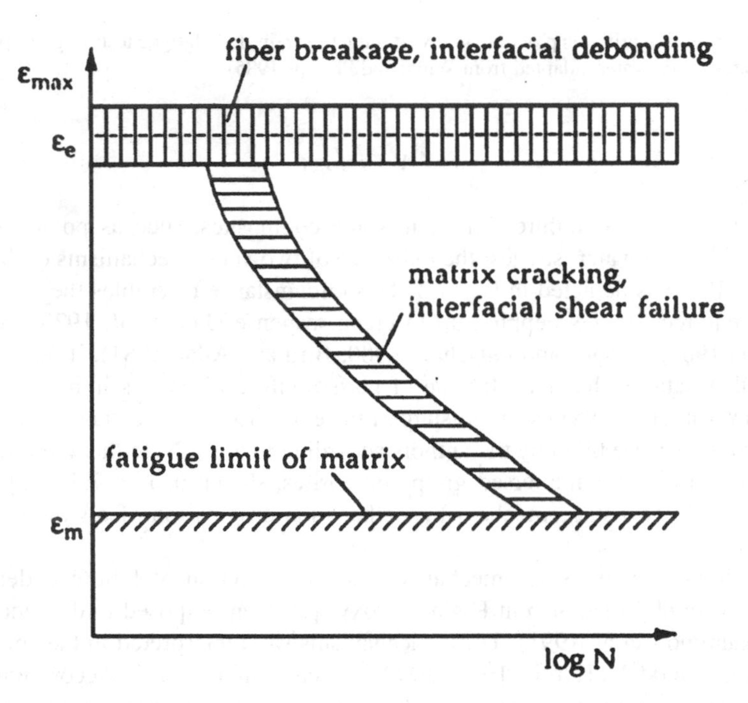

Fig. 7.5 Fatigue-life diagram for unidirectional composites under loading parallel to fibers (adapted from Talreja 1981)

For glass fiber composites, static fatigue limits correspond to about 60% of their "instantaneous" static, strength (Thomas 1960; Metcalfe and Schmitz 1972; Roberts 1978; Aveston et al. 1980; Chateauminois et al. 1993). However, significantly larger reductions in both crack initiation times and failure times were noted in several glass/polymer systems immersed in a solution of sulfuric acid (van den

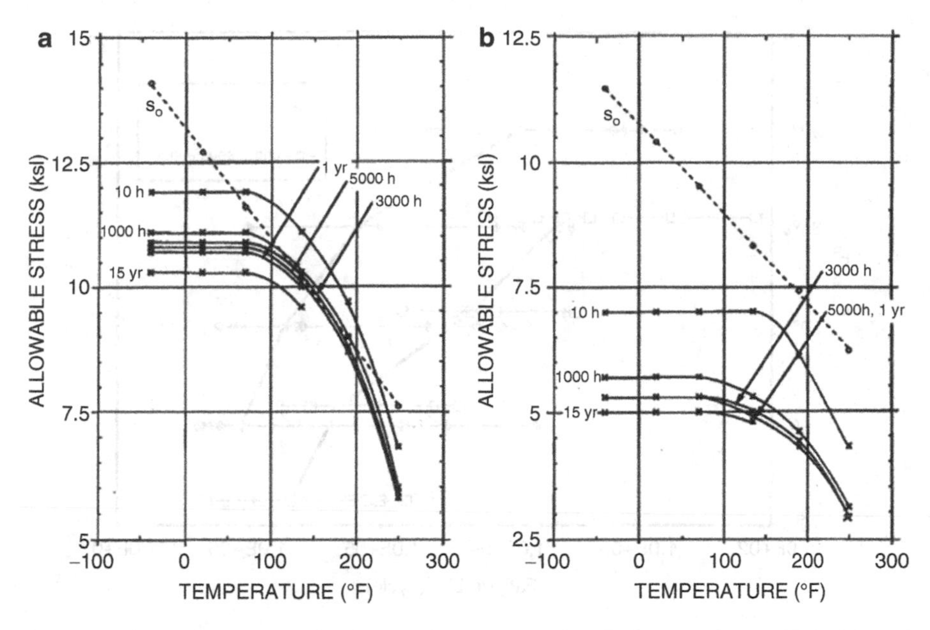

Fig. 7.6 Allowable levels of stress for swirl-mat, glass fiber/urethane composites for various durations of sustained loading vs. temperature (**a**) dry; (**b**) immersed in distilled water. Note the reduction, by about 40%, attributed to distilled water (after Corum et al. 1998)

Emde and van den Dolder 1991). Obviously, the durability of polymeric composites exposed to solvents is affected by temperature (Charles 1958b), since it is governed by thermally activated processes.

In the case of glass fiber cloth reinforced by vinyl ester resins that were immersed in hot water (Morii et al. 1991), it was possible to associate the degradation process with fiber/matrix interfacial failures. This was achieved by considering distinct fiber surface treatments and observing the existence of good correlations between the "knee-point" stress levels in stress-strain plots and water weight gain data.

Chopped mat glass fiber composites exhibit weight losses when exposed to hot water above 90°C. These weight losses correlate with reduction of up to 50% in strength and stiffness (Lou and Murtha 1988). Note again that 90°C is unlikely to apply under realistic circumstances.

An extensive program concerning the behavior of urethane matrix/glass fiber swirl mat composites indicated that distilled water has the most severe effects of all automotive fluids on the strength and durability of those composites (Corum et al. 1998). These observations necessitated a 40% knock-down factor in short-term stress allowable and a 50% reduction for the long-term allowable design stress. Higher knock-down factors were recommended for elevated temperatures. The respective diagrams are shown in Figs. 7.6a and 7.6b (Corum et al. 1998).

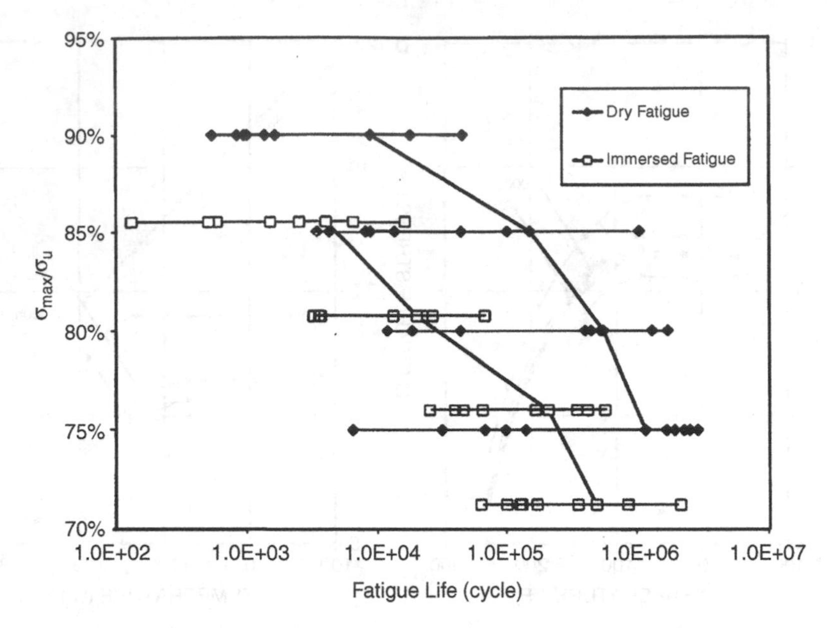

Fig. 7.7 *S–N* plots for $[0°/90°_3]_s$ gr/ep coupons subjected to dry and (presaturated) immersed fatigue

Basic considerations of crack growth at various stages of the stress corrosion cracking process (Evans 1972) suggest that chemical reactions control crack growth at low-load levels while diffusion of corrosive substance governs the failure process under higher loads. At yet higher stress levels, crack growth is dominated by stress-assisted corrosion.

Upon drying, composites regain a portion or the entire value of their original strength (Phillips et al. 1978; Dewimille et al. 1980; Springer et al. 1981; Manocha et al. 1982; Drzal et al. 1985). The amount of recovery depends on the extent and kind of irreversible damage caused by the prior exposure to solvents. Generally speaking, a nearly complete recovery of properties is associated with sorption along curves "LF," "A," and "B" in Fig. 4.2 while permanent property losses are noted for weight gain data along curve "C" and "D" in that figure.

Prestressing of composites prior to their exposure to fluids may accelerate the solvent-induced corrosion (Sandifer 1982; Fujii et al. 1993). This enhancement is most likely due to mechanical damage caused by prestressing, which facilitates the subsequent penetration of fluid through capillary action (Ehrenstein and Spaude 1984).

Though generalizations are risky and unreliable, it seems that in many circumstances, water (distilled or salty) causes more significant reductions in the strength of composites than fuels or motor oils (Sandifer 1982; Blicblau et al. 1993; Corum et al. 1998). In some cases, an increase in water salinity tends to intensify the losses in composites' properties (Nakanishi and Shindo 1982).

The presence of fluids appears to have only a marginal effect on impact damage of polymeric composites as well as on compressive strength after impact. This is due to the fact that the increased overall wet compliance of the composite laminate tends to localize the damage zone (Saito and Kimpara 2007; Imielinska and Guillaumat 2004). Nevertheless, it is highly risky to generalize the above results, since data may vary substantially with material, lay-up, impact energy, and geometrical configurations.

7.2 Fluid Effects on the Fracture Toughness, Fatigue Response, and Impact Resistance of Polymeric Composites

From the microlevel point of view, the fracture toughness of composite materials depends on the combined interplay of energy-absorbing micromechanisms in the vicinity of a crack tip (Jordan 1985). These mechanisms are governed by the ductilities of the resin and the interphase material, as well as of the fiber/matrix interfacial bond. It is important to recognize that the capacity of the resin and the interphase matter to undergo ductile deformation is restricted by the confining effects of the stiffer and more brittle fibers. Consequently, substantial improvements in the toughness of resins may result in only marginal increases in the toughness of composites.

Liquids enhance the ductility of polymers through plasticization, which should tend to increase the toughness of composites. However, liquids may act to weaken the fiber/matrix interfacial bond and induce osmotic pressures which would augment in the activation of crack growth, thereby diminishing the toughness of composites. The interplay among these competing effects is difficult to quantify and may be the underlying reason for the contradictory information reported in the literature. For instance, these contradictory effects may cancel each other, as evidenced by the insensitivity of the extent of edge delaminations in cross-ply glass/epoxy laminates to the presence or absence of water (Yang et al. 1992), and of the delamination fracture toughness reported for unidirectionally reinforced gr/ep coupons (Russell and Street 1985). Similar insensitivity was observed in the growth of edge delamination in cross-ply gr/ep laminates immersed in sea water, in spite of some differences in detail when compared to dry circumstances (Chiou and Bradley 1993). On the other hand, it was demonstrated that in gr/ep laminates, the ingress of fluids results in a monotonic reduction in the required energy for delamination with increasing amounts of absorbed fluid (Hooper et al. 1991b). Similarly, the levels of both initiation and critical energies for mode II fracture decreased upon the uptake of fluids in glass/polypropelene composites. The latter decrease was noted under exposures to distilled water as well as to sea water (Davies et al. 1996a, b). Also, a decrease in mode II critical energy, G_{IIC}, was noted for gr/ep coupons exposed to various levels of relative humidity (Zhao and Gaedke 1996). However, in contrast to the latter observations, the critical energy required to initiate delamination in gr/ep composites increased with the amount of absorbed fluid (Hooper et al. 1991a).

Exposure of unidirectionally reinforced E glass plastics to the combined effects of environment and stress may lead to their dramatic embrittlement (Price 1989). On the other hand, the presence of liquids may greatly enhance the fracture toughness of other composites in certain circumstances, as occurs for mode I delamination of unidirectionally reinforced gr/ep double-cantilever beam specimens immersed in sea water (Sloan and Seymour 1992; Walker and Hu 2003; Czigány et al. 1996; Ishak et al. 2000). The reason for the latter, seemingly surprising, result is that the extensive weakening of the fiber–matrix interfaces caused by sea water gives rise to fiber bridging across the crack surfaces.

Finally, it is worth noting that at least some of the apparent disagreement regarding fluid effects on fracture toughness may be attributable to the utilization of various lay-ups in the reported experimental programs, e.g., coupons containing plies oriented at angles $\pm\,\theta$ (O'Brien et al. 1986) as well as the resin material (Johannesson and Blikstad 1985).

As a general trend, fiber bridging diminishes when delaminations occur between layers of larger disparities in the directions of fiber reinforcement. In such circumstances, the delamination fracture toughness of the dry case is likely to exceed its wet value.

Fatigue response is characterized by plotting crack velocity (da/dt) vs. stress intensity (K_I) (Friedrich and Karger-Kocsis 1990) or by the familiar *S–N* diagrams, except that composites exhibit larger data scatter than metals. The *S–N* data can be related by empirical expressions, with attempted predictions and reliability assessments based upon statistical considerations (Sendeckyj 1990; Talreja 1981, 1990).

In a variety of circumstances, it was observed that fluids accelerate the fatigue failure process and shorten the fatigue life of glass fiber composites (Boller 1964; Phillips et al. 1978; Mandell 1979; Aveston et al. 1980; Dewimille et al. 1980; Lou and Murtha 1988; Friedrich and Karger-Kocsis 1990; Yang et al. 1992) and carbon fiber composites (Sumsion 1976; Morton et al. 1988). That observed reduction in fatigue life is by no means universal since no discernible reduction was noted for a quasi-isotropic lay-up of S2 glass/3M composite at several levels of relative humidity (Zaffaroni 1997). It must, therefore, be surmised that the extent of these effects depends on both fluid type and material system (Sandifer 1982). In addition, the presence of water may affect the failure mechanisms that evolve during the fatigue process (Morton et al. 1988).

It is reasonable to assume that the above-mentioned shortening in fatigue life is caused by the same fluid-induced mechanisms that degrade the strength of composites. Nevertheless, there exist two additional degradation mechanisms that are specific to fatigue. The first involves the synergistic interaction between the solvent sorption process and fatigue-induced damage, namely a profusion of microcracks that open capillary paths for solvent ingress (Jones et al. 1984). As noted in Sect. 4.10 and shown in Fig. 4.31, this mode of solvent penetration is orders of magnitudes faster than the diffusion process and subjects the composite to accelerated degradation (Kosuri and Weitsman 1995). The second mechanism involves the synergistic destruction of interfacial fiber–matrix bonds caused by their cyclic rubbing under fatigue loading. The introduction of liquids into the

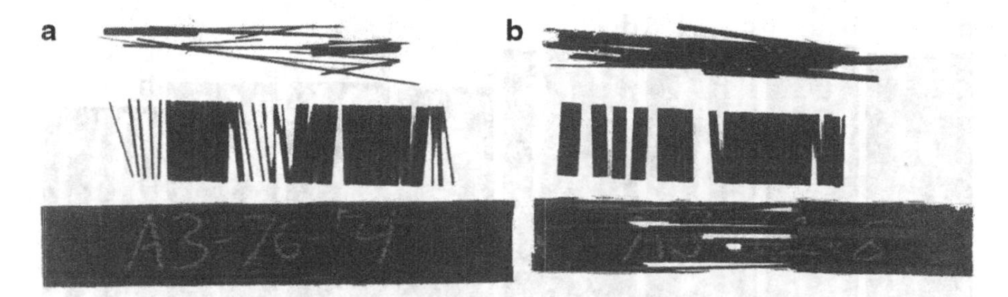

Fig. 7.8 Fatigue failure in $[0°/90°_3]_s$ AS4/3501-6 coupons under (**a**) dry, and (**b**) presaturated immersed circumstances. $R = \sigma_{max}/\sigma_{uts} = 0.80$ and $R = \sigma_{max}/\sigma_{uts} = 0.3$. Note the relatively fewer delaminations and the higher density of transverse cracks in the dry case

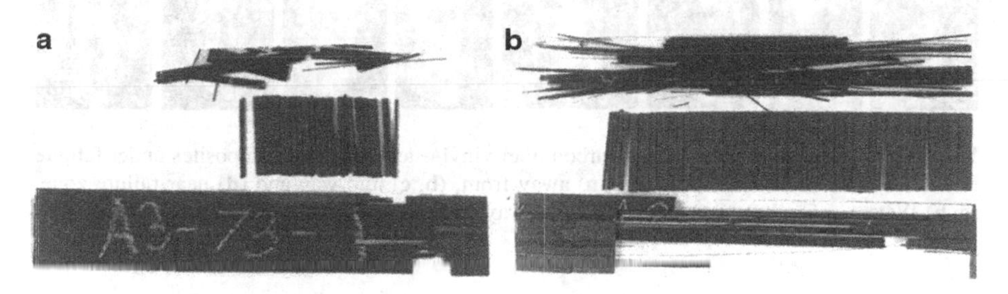

Fig. 7.9 Same as Fig. 8.6 but $\sigma_{max}/\sigma_{uts} = 0.90$ and $R = \sigma_{max}/\sigma_{uts} = 0.1$

composite enhances the above phenomenon, since liquids act as lubricants (Jones et al. 1984). Evidence for the smoothening and polishing of fiber–matrix interfaces due to moisture implies that the same effect occurs during fatigue (Aveston and Sillwood 1982; Galea and Saunders 1993).

All the "wet fatigue" data noted thus far were obtained for specimens with prior, but not current, exposure to fluids. However, several studies (Kosuri and Weitsman 1995; Gao and Weitsman 1998) on the effects of sea water on the fatigue of cross-ply gr/ep composites indicate that significant reductions in fatigue life occur especially under immersed conditions, when capillary action is highly accentuated. The resulting *S–N* curves for cross-ply gr/ep laminates are shown in Fig. 7.7. The reduced fatigue life under immersed conditions is observed to occur in association with profound interplay delaminations, as shown in Figs. 7.8 and 7.9 for graphite/epoxy laminates and recently reproduced for $[0°/90°]_{2s}$ piecewise stitched composite facing in Fig. 7.10 (Weitsman et al. 2009 and Siriruk et al. 2010). The results are shown in Figs. 7.11 and 7.12. Analytical models that explain this phenomenon attributed the extensive delaminations observed under immersed fatigue to the presence of nearly incompressible water within the fatigue-induced microcracks. These models suggest that water that enters those microcracks by capillary motion cannot escape into the ambient environment during the downloading stages of the fatigue cycle and is compressed into the interply delaminations as sketched in Fig. 7.13 (Smith and

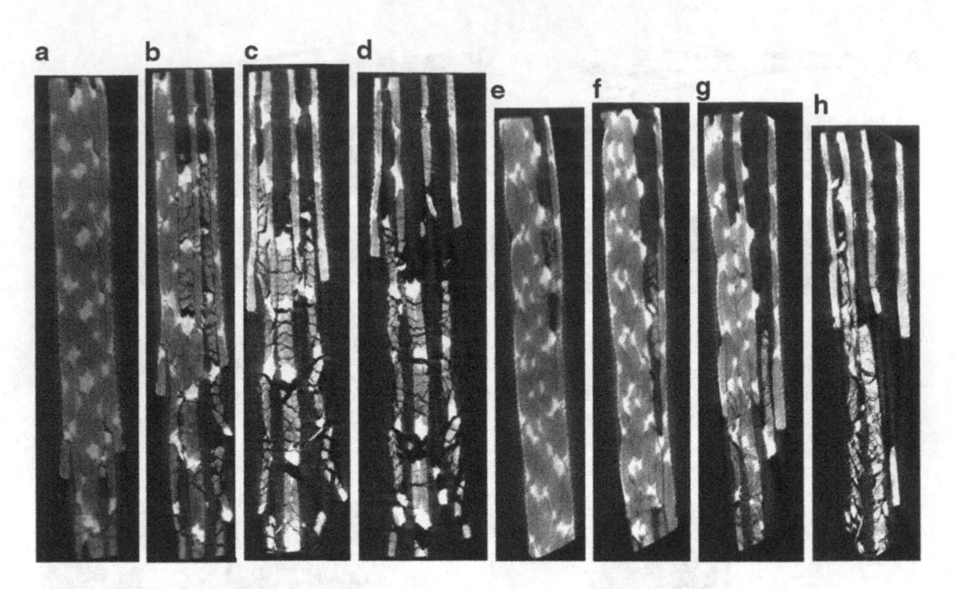

Fig. 7.10 X-ray CT results of failed carbon fiber vinyl/ester laminated composites under fatigue test. (**a–d**) Unconfined dry specimen, (**a**) away from, (**b**, **c**) mid way and (**d**) near failure zone. (**e–h**) Water confined and wet specimens, (**e**) away from, (**f**, **g**) mid way and (**h**) near failure

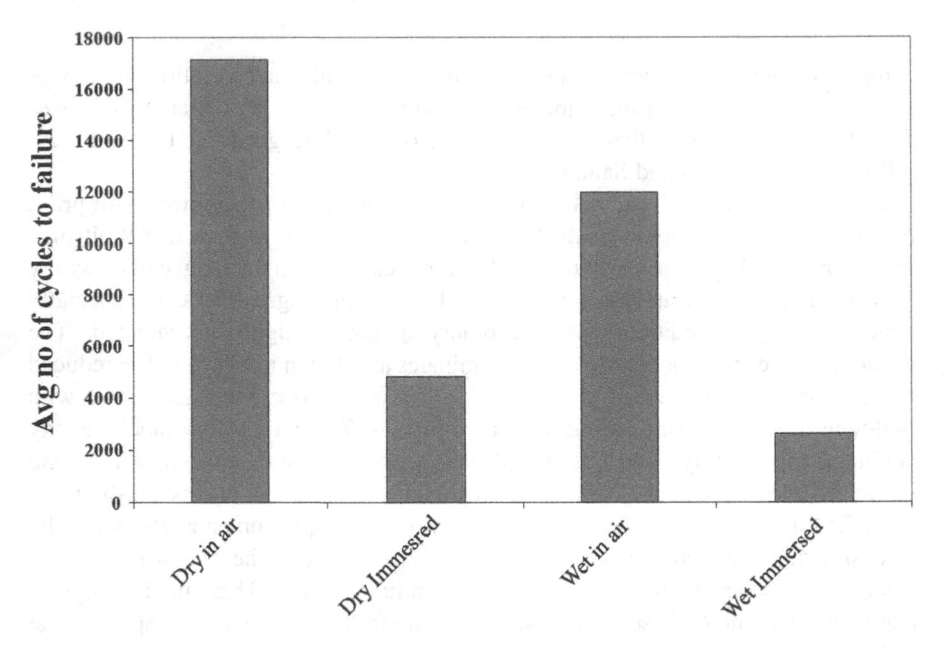

Fig. 7.11 Average number of cycles to failure of dry and wet $[\pm45]_{2S}$ sample in air and under immersed conditions under cyclic loading at 1 Hz frequency

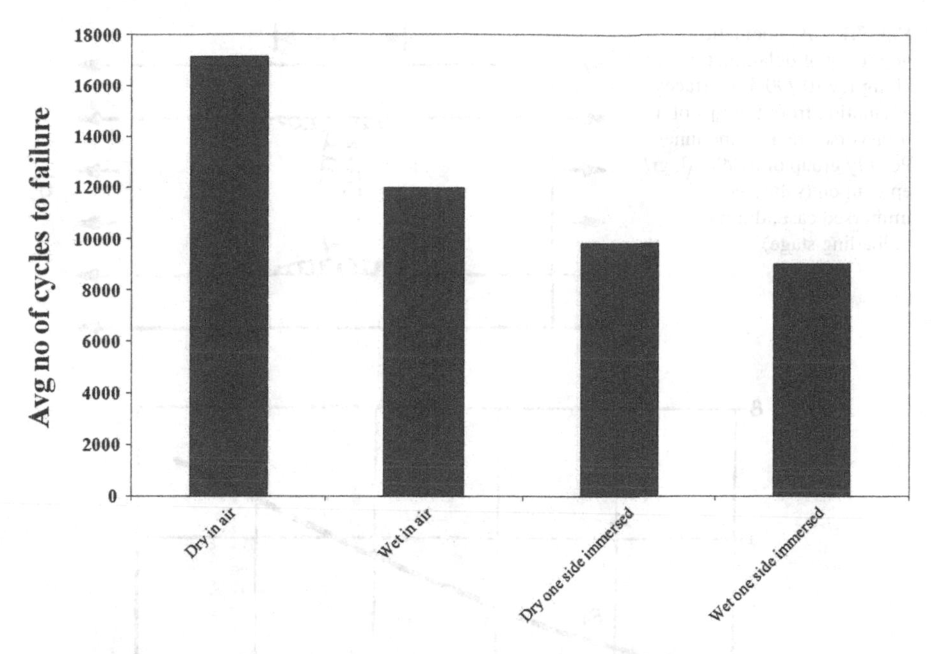

Fig. 7.13 Average number of cycles to failure of dry and wet $[\pm 45]_{2S}$ sample in air and under one sided exposure conditions under cyclic loading at 1 Hz frequency

Weitsman 1996). In this manner, delamination, rather than additional transverse cracking, becomes the preferred failure mode. Results for comparative energy levels for dry and immersed fatigue are shown in Fig. 7.14 (Selvarathinam and Weitsman 1998). Note that growths of the delaminations alluded to in Fig. 7.14 occur when the energies available for their propagation exceed the relative fracture energies (G/G_c) for further transverse cracking. The transition from transverse cracking to delamination occurs at crack densities of (a) 0.52 mm^{-1}, (b) 0.89 mm^{-1}, and (c) 0.9 mm^{-1} for the SI (saturated immersed), DIA (dry-in-air), and SIA (saturated-in-air) fatigue loadings. The dashed lines in Fig. 7.14, which correspond to (G/G_c) < 1, indicate delamination arrest. The results shown in Fig. 7.7 can therefore be rationally explained by fracture mechanics data exhibited in Figs. 7.8–7.12 and the ensuing analyses. Subsequent experiments showed that even more dramatic reductions in fatigue life under immersed conditions occur in cross-ply glass/epoxy laminations, as shown in Fig. 7.15 (Gao and Weitsman 1998). Similar results were obtained for glass/vinyl ester composites (Hayes et al. 1998) and glass chopped strand mat/polyester composites (Hasan et al. 1998). In that case, it is interesting to note that no distinct ply-level failure mechanisms can be discerned between dry and immersed fatigue and the only observable difference is noted on the microlevel. A substantial amount of pitting is seen on the failure surfaces of the immersed coupons, which is absent in the dry case, as shown in Fig. 7.16a, b. It was also noted, for another glass fiber system (Hayes et al.), that while the *S–N* curves for the dry and immersed fatigue of glass fiber composites remained widely separated when plotted vs. the absolute value of the

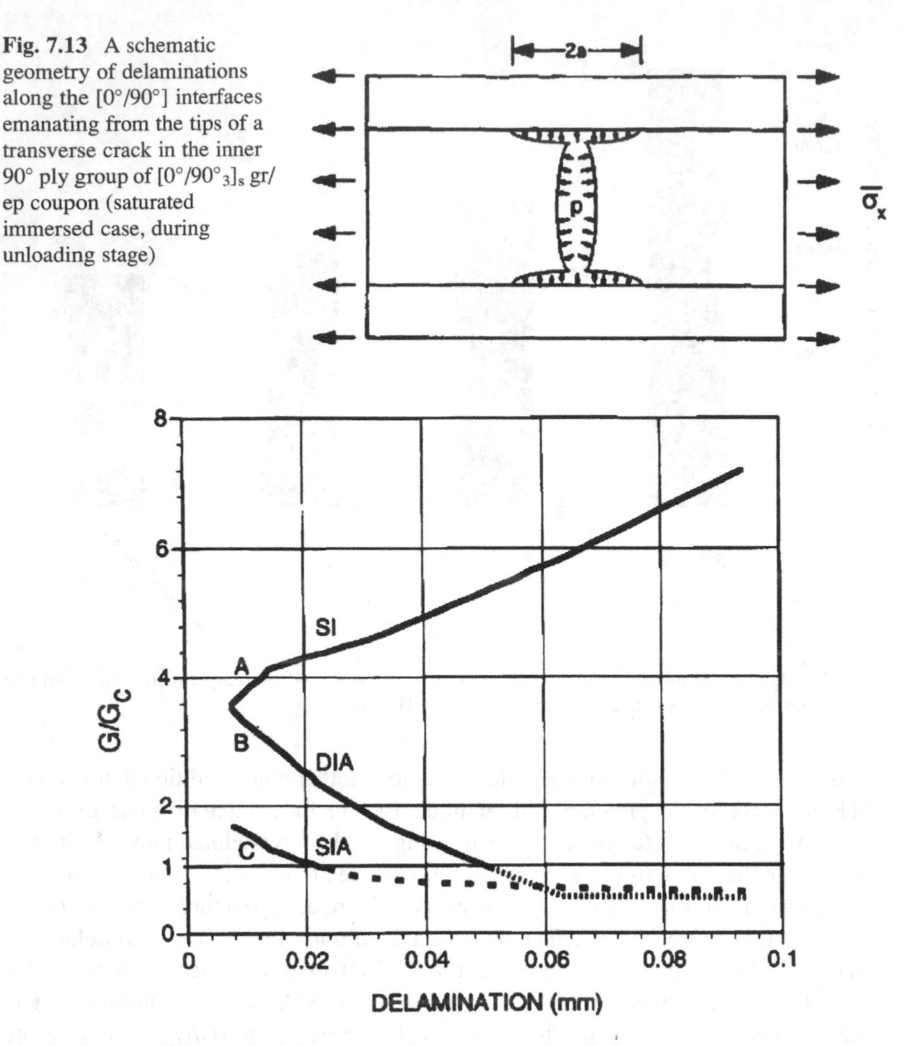

Fig. 7.13 A schematic geometry of delaminations along the [0°/90°] interfaces emanating from the tips of a transverse crack in the inner 90° ply group of $[0°/90°_3]_s$ gr/ep coupon (saturated immersed case, during unloading stage)

Fig. 7.14 The ratio G/G_c of available to required levels of fracture energies for the propagation of delamination along the [0°/90°] interface in $[0°/90°_3]_s$ AS4/3501-6 coupons subjected to fatigue. The curve SI, DIA, and SIA correspond to the saturated-immersed, dry-in-air, and saturated-in-air case, respectively

stress S, these curves coincide when plotted against the common ordinate $S/S_{\text{ult.dry}}$ and $S/S_{\text{ult.saturated}}$. In other words, immersed and dry fatigue data can be scaled by the factor $S_{\text{ult.dry}}/S_{\text{ult.saturated}}$. Such an observation would suggest that for glass fiber composites, damage due to fluids is separable from fatigue-induced damage. Such a suggestion was put forward elsewhere as well (Zaffaroni 1997)

Though it may be argued that the aforementioned results are unrealistically amplified by fluid influx across the exposed edges of the test samples, a similar reduction in fatigue life is implied by the damage vs cyles-to-failure in cylindrical

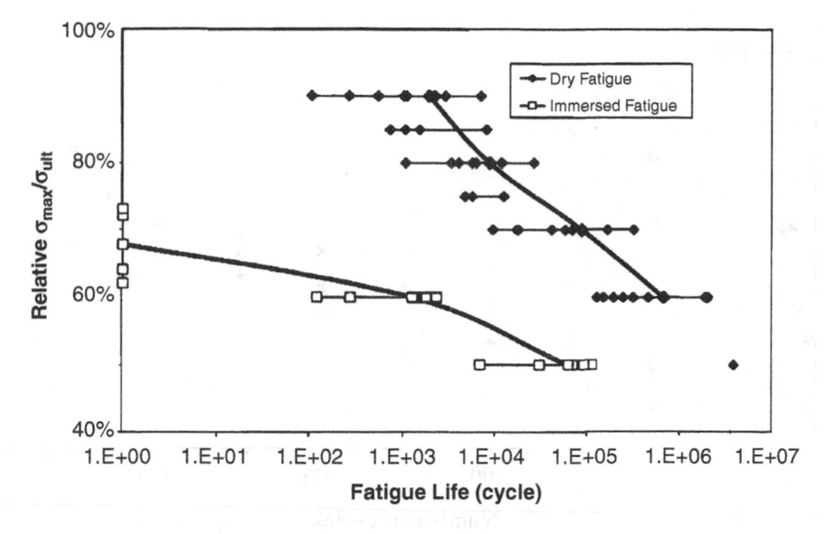

Fig. 7.15 *S–N* curve under dry and immersed fatigue of $[0°/90°_3]_s$ E-glass/SP1003 coupons with $R = \sigma_{max}/\sigma_{min} = 0.1$

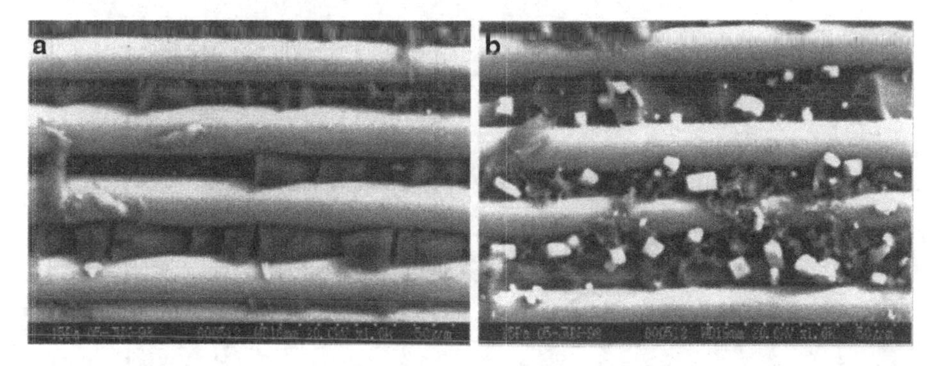

Fig. 7.16 SEM observation of failed $[0°/90°_3]_s$ glass/epoxy coupons under (**a**) dry and (**b**) immersed fatigue. Note the evident presence of mode II fracture component in dry case, as contrasted with the pitting and pulling in the immersed circumstance. The *white blocks* in (**b**) are salt crystals

specimens, as shown in Fig. 7.17 (Perreux and Suri 1997). Also, it is worth noting that *S–N* plots resembling Fig. 7.6 were obtained more recently for plain woven and multiaxial knitted CFRP laminates subjected to dry and "wet-in-air" fatigue in Fig. 7.18 (Kimpara and Saito 2007).

The effects of fluids on the impact resistance of composites contain the same contradictory aspects that hold for fracture toughness. In view of the solvent-induced weakening of the fiber/matrix interfacial strength, the impacting object may engage a larger volume of fibers in the absorption of the impact energy, thus resulting in improved impact resistance (Strait et al. 1992; Lin 1993). In addition,

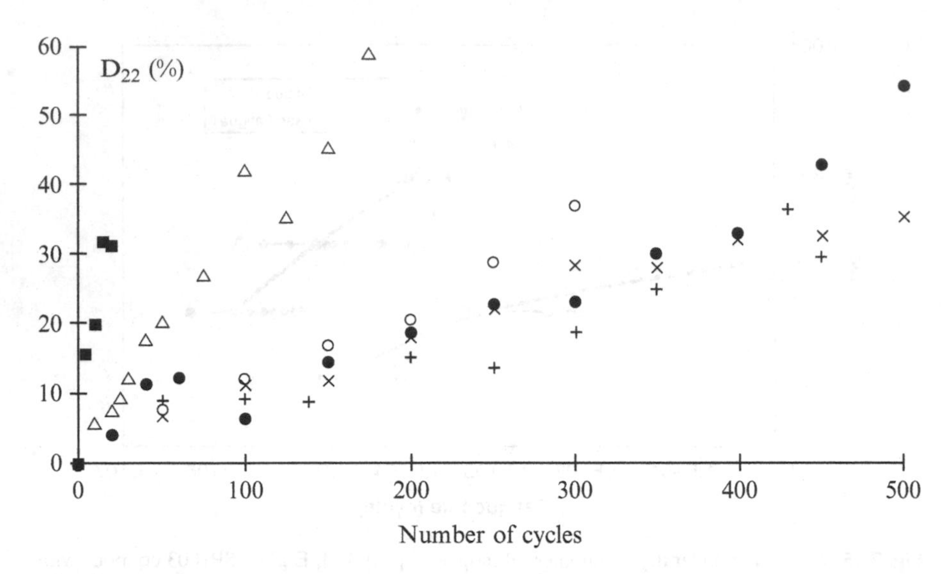

Fig. 7.17 Influence of water content on the fatigue damage kinetics: *plus or minus*, oven dried; *multiplication sign*, unaged; *open diamond*, aged for 15 days; *open inverted triangle*, aged for 2 months; *filled square*, aged for 18 months; *filled circle*, aged for 18 months and oven-dried again

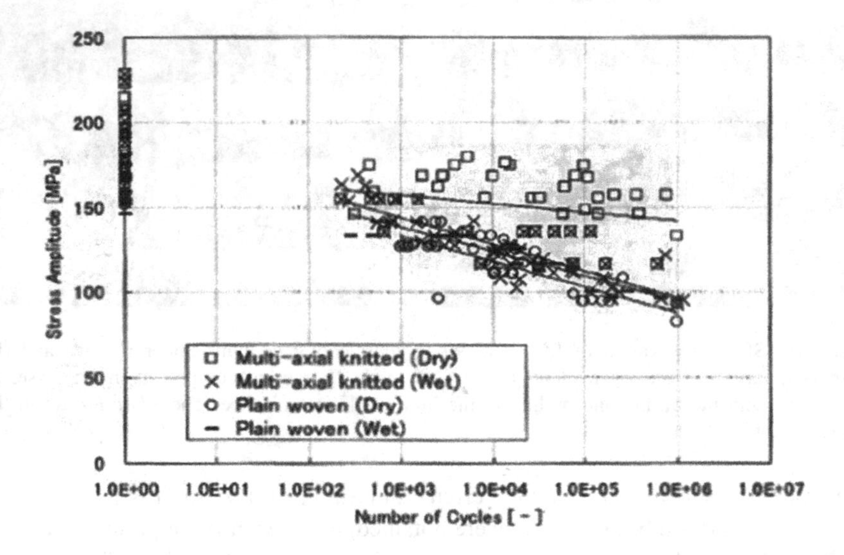

Fig. 7.18 Fatigue strength under water absorption condition of plain woven and multi-axial knitted CFRP laminates (Kimpara and Saito 2007)

the plasticization and enhanced ductility of the wet polymer may explain the observed reduction in the peak amplitudes of the contact pressures during impact (Strait et al. 1992; Lin 1993). On the other hand, impact-induced damage may enhance the amounts of absorbed fluids within the composite under subsequent ambient exposures and lead to long-term degradation in residual mechanical properties (Ma et al. 1991).

References

Ashbee KHG, Wyatt RC (1969) Water damage in glass fibre/resin composites. Proc R Soc Lond A Math Phys Sci 312(1511):553–564

Aveston J, Sillwood JM (1982) Long-term strength of glass-reinforced plastics in dilute sulphuric acid. J Mater Sci 17(12):3491–3498

Aveston J, Kelly A, Sillwood JM (1980) Long term strength of glass reinforced plastics in wet environments. In: Bunsell AR, Bathias C, Martrenchar A, Menkes D, Verchery G (eds) Proceedings of the third international conference on composite materials (ICCM/3), Paris, 26–29 August 1980, pp 556–568

Bailey JE, Fryer TMW, Jones FR (1980) Environmental stress-corrosion edge-cracking of glass reinforced polyesters. In: Bunsell AR, Bathias C, Martrenchar A, Menkes D, Verchery G (eds) Proceedings of the third international conference on composite materials (ICCM/3), vol 1: Advances in composite materials, Paris, 26–29 August, pp 513–528

Bird J, Allan RC (1981) The development of improved FRP laminates for ship hull construction. In: Marshall IH (ed) Composite structure. Applied Science, New York, pp 202–223

Blicblau AS, Tran L, Warden P (1993) Environmental effects on carbon fibre/epoxy shear strength. In: Miravete A (ed) Proceedings of the ninth international conference on composite materials (ICCMIS), vol 5: Composite behaviour, Madrid, 12–16 July 1993, pp 660–664

Boller KH (1964) Fatigue characteristics of RP laminates subjected to axial loading. Mod Plast 188:145–150

Browning CE, Husman GE, Whitney JM (1976) Moisture effects in epoxy matrix composites. Composite materials: testing and design, ASTM STP 617, Philadelphia, pp 481–496

Castaing PH, Lemoine L, Tsoubalis N (1993) Experimental study of the variation in mechanical characteristics of orthotropic laminates immersed in water by a non destructive method. In: Miravete A (ed) Proceedings of the ninth international conference on composite materials (ICCMIS), vol 5: Composite behaviour, Madrid, 12–16 July 1993, pp 577–584

Charles RJ (1958a) Static fatigue of glass. I. J Appl Phys 29(11):1549–1553

Charles RJ (1958b) Static fatigue of glass. II. J Appl Phys 29(11):1554–1560

Chateauminois A, Chabert B, Soulier JP, Vincent L (1993) Effects of hygrothermal aging on the durability of glass/epoxy composites. Physico-chemical analysis and damage mapping in static fatigue. In: Miravete A (ed) Proceedings of the ninth international conference on composite materials (lCCM/9), vol 5, Madrid, pp 593–600

Chiou P, Bradley WL (1993) The effect of sea water exposure on the fatigue edge delamination growth of a carbon/epoxy composite. In: Miravete A (ed) Proceedings of the ninth international conference on composite materials (lCCM/9), vol 5, Madrid, pp 516–523

Corum JM, Battiste RL, Brinckman CR, Ren W, Ruggles MB, Yahr GT (1998) Durability-based design criteria for an automotive structural composite: part I. Design rules. Oak Ridge National Laboratory, Oak Ridge

Czigány T, Ishak ZAM, Heitz T, Karger-Kocsis J (1996) Effects of hygrothermal aging on the fracture and failure behavior in short glass fiber-reinforced, toughened poly(butylene terephthalate) composites. Polym Compos 17(6):900–909

Davies P, Choqueuse D, Mazeas F (1998) Composites underwater. In: Reifsnider KL, Dillard DA, Cardon AH (eds) Progress in durability analysis of composite systems. Balkema, Rotterdam, pp 19–24

Davies P, Pomies F, Carlsson LA (1996a) Influence of water absorption on transverse tensile properties and shear fracture toughness of glass/polypropylene. J Compos Mater 30 (9):1004–1019

Davies P, Pomiès F, Carlsson LA (1996b) Influence of water and accelerated aging on the shear fracture properties of glass/epoxy composite. Appl Compos Mater 3(2):71–87

Dewimille B, Thoris J, Mailfert R, Bunsell AR (1980) Hydrothermal aging of an unidirectional glass-fibre epoxy composite during water immersion. In: Bunsell AR (ed) Advances in composites materials; Proceedings of the third international conference on composite materials, Paris, France. Pergamon Press, Oxford, pp 597–612

Drzal LT, Rich MJ, Koenig MF (1985) Adhesion of graphite fibers to epoxy matrices. III. The effect of hygrothermal exposure. J Adhes 18:49–72

Ehrenstein GW, Spaude R (1984) A study of the corrosion resistance of glass fibre reinforced polymers. Compos Struct 2:191–200

Evans AG (1972) A method for evaluating the time-dependent failure characteristics of brittle materials – and its application to polycrystalline alumina. J Mater Sci 7:1137–1146

French MA, Pritchard G (1991) Strength retention of glass/carbon hybrid laminates in aqueous media. In: Cardon AH, Verchery G (eds) Durability of polymer based composite systems for structural applications. Elsevier Applied Science, New York, pp 345–354

Friedrich K (1981) Stress corrosion crack propagation in glass fibre reinforced/thermoplastic PET. J Mater Sci 16(12):3292–3302

Friedrich K, Karger-Kocsis J (1990) Fracture and fatigue of unfilled and reinforced polyamides and polyesters. In: Schults JM, Fakirov S (eds) Solid state behavior of linear polyesters and polyamides. Prentice-Hall, Englewood Cliffs, pp 249–322

Fujii Y, Maekawa Z, Hamada H, Tubota T, Murakami A, Yoshiki T (1993) Evaluation of initial damage and stress corrosion of GFRP. In: Miravete A (ed) Proceedings of the ninth international conference on composite materials (ICCMIS), vol 5: Composite behaviour, Madrid, 12–16 July 1993, pp 562–568

Galea SC, Saunders DS (1993) Effects of hot/wet environments on the fatigue behaviour of composite-to-metal mechanically fastened joints. In: Chandra T, Dhingra AK (eds) Proceedings of the international conference on advanced composite materials (ICACM), Advanced composites '93, Wollongong, Australia, 15–19 February 1993, pp 525–531

Gao J, Weitsman YJ (1998) Composites in sea water: sorption, strength and fatigue, University of Tennessee Report MAES 98–4.0 CM, August 1998

Grant TS (1991) Sea water degradation of polymeric composites. Masters Thesis, Texas A & M University

Hamada H, Maekawa Z, Morii T, Gotoh A, Tanimoto T (1991) Durability of adhesive bonded FRP joints immersed in hot water. In: Cardon AH, Verchery G (eds) Durability of polymer based composite systems for structural applications, vol 46. Elsevier, Amsterdam, pp 418–427

Haque A, Mahmood S, Walker L, Jeelani S (1991) Moisture and temperature induced degradation in tensile properties of kevlar-graphite/epoxy hybrid composites. J Reinforced Plast Compos 10(2):132–145

Hasan EH, Abdel-Hakeeln HM, El-Sayed AAM (1998) Environmental rotating bending fatigue behavior of fiber glass/polyester composites. In: Crivelli-Visconti I (ed) Proceedings of ECCM-8 European conference on composite materials, vol 3, Naples, Italy. Woodhead Publishing, Cambridge, pp 91–198

Hayes MD, Garcia K, Verghese N, Lesko JJ (1998) The effects of moisture on the fatigue behavior of glass/vinyl ester composite. In: Saadatmanesh H, Ehsani MR (eds) Proceedings of the second international conference on composites in infrastructure (ICCI '98), Tucson, Arizona, USA, 5–7 January 1998, pp 1–13

Hertz J (1973) Investigation into the high-temperature strength degradation of fiber-reinforced resin composite during ambient aging. Report No. GDCA-DBG73-005, Contract NAS8-27435 (June 1973), Convair Aerospace Division, General Dynamics Corporation

Hofer Jr KE, Larsen D, Humphreys VE (1975) Development of engineering data on the mechanical and physical properties of advanced composites materials. Technical Report AFML-TR-74-266 (February 1975), Air Force Materials Laboratory, Air Force Systems Command, Wright-Patterson Air Force Base, Dayton

Hofer KE Jr, Rao N, Larsen D (1974) Development of engineering data on the mechanical and physical properties of advanced composite materials, Technical Report AFML-TR-72-205, Part II (February 1974). Air Force Materials Laboratory, Air Force Systems Command, Wright-Patterson Air Force Base, Dayton

Hogg PJ, Hull D (1982) Role of matrix properties on the stress corrosion of GRP. 13th reinforced plastics congress, Brighton, The British plastics federation, London, pp 115–120

Hogg PJ, Hull D, Legg MJ (1981) Failure of GRP in corrosive environments. In: Marshall IH (ed) Composites structures. Applied Science Publishers, New York, pp 106–122

Hooper SJ, Subramanian R, Toubia RF (1991a) Effects of moisture absorption on edge delamination, Part II: an experimental study of jet fuel absorption on graphite epoxy. ASTM STP 1110:107–125

Hooper SJ, Toubia RF, Subramanian R (1991b) Effects of moisture absorption on edge delamination, Part I: analysis of the effects of nonuniform moisture distributions on strain energy release rates. ASTM STP 1110:89–106

Hsu P-L, Chou T-W (1985) Corrosion effects on e-glass filaments, bundles, and their aligned short-fiber composites. In: Harrigan Jr WC, Strife J, Dhingra AK (eds) Proceedings of the fifth international conference on composite materials (ICCM/5), San Diego, 29 July–1 August 1985, pp 1475–1490

Husman GE (1976) Characterization of wet composite materials. Presented at the mechanics of composites review, Bergamo Center, Dayton, 28–29 January 1976

Imielinska K, Guillaumat L (2004) The effect of water immersion ageing on low-velocity impact behaviour of woven aramid-glass fibre/epoxy composites. Compos Sci Technol 64 (13–14):2271–2278

Ishak ZAM, Ishiaku US, Karger-Kocsis J (2000) Hygrothermal aging and fracture behavior of short-glass-fiber-reinforced rubber-toughened poly(butylene terephthalate) composites. Compos Sci Technol 60(6):803–815

Jain RK, Asthana KK (1980) Effect of natural weathering on the creep behavior of GRP laminates in tropical climates. In: Bunsell AR, Bathias C, Martrenchar A, Menkes D, Verchery G (eds) Proceedings of the third international conference on composite materials (ICCM/3), vol 1: Advances in composite materials, Paris, 26–29 August, pp 613–623

Jain RK, Goswamy SK, Asthana KK (1979) A study of the effect of natural weathering on the creep behaviour of glass fibre-reinforced polyester laminates. Composites 10(1):39–43

Johannesson T, Blikstad M (1985) Influence of moisture and resin ductility on delamination. Compos Sci Technol 24(1):33–46

Jones CJ, Dickson RF, Adam T, Reiter H, Harris B (1984) The environmental fatigue behaviour of reinforced plastics. Proc R Soc Lond A 396(1811):315–338

Jones FR, Rock JW, Wheatley AR, Bailey JE (1982) The environmental stress corrosion cracking of glass fibre reinforced polyester and epoxy composites. In: Hayashi T, Kawata K, Umekawa S (eds) Proceedings of the fourth international conference on composite materials (ICCM/4), vol 2: Progress in science and engineering of composites, Tokyo, 25–28 October 1982, pp 929–936

Jordan WM (1985) The effect of resin toughness on the delamination fracture behaviour of graphite/epoxy composites. Ph.D. Dissertation, Interdisciplinary Engineering, Texas A & M University, December 1985

Juska T (1993) Effect of water immersion on fiber/matrix adhesion. CarderockDiv-SME-92/38, Ship Materials Engineering Department Research and Development Report, US Navy, Carderock Division, Naval Surface Warfare Center, Bethesda (January 1993)

Kaelble DH, Dynes PJ, Crane LW, Maus L (1975) Interfacial mechanisms of moisture degradation in graphite-epoxy composites. J Adhes 7(1):25–54

Kaminski BE (1973) Effects of specimen geometry on the strength of composite materials. Analysis of the Test Methods for High Modulus Fibers and Composites, ASTM STP 521, pp 181–191

Kelly A, McCartney LN (1981) Failure by stress corrosion of bundles of fibres. Proc R Soc Lond A 374(1758):475–489

Kimpara I, Saito H (2007) Part II: Post-impact fatigue behavior of woven and knitted fabric CFRP laminates for marine use. Long-term durability and damage tolerance of innovative marine composites (NICOP), Technical Reports and Data Base for Office of Naval Research Project, Feb 2007

Kosuri R, Weitsman YJ (1995) Sorption processes and immersed fatigue response of gr/ep composites in sea water. In: Poursartip A, Street K (eds) Proceedings of the Tenth International Conference on Composite Materials (lCCM-10), vol 4, Whistler, BC, pp 177–184

Kriz RD, Stinchcomb WW (1982) Effects of moisture, residual thermal curing stresses, and mechanical load on the damage development in quasi-isotropic laminates. In: Reifsnider KL (ed) Damage in composite materials, ASTM STP 775. American Society for Testing and Materials, Philadelphia, pp 63–80

Lin C-W (1993) Effect of moisture absorption on the impact behavior of unidirectional carbon fiber reinforced nylon 6 composite. In: Chandra T, Dhingra AK (eds) Proceedings of the international conference on advanced composite materials (ICACM), Advanced composites '93, Wollongong, Australia, 15–19 February 1993, pp 597–602

Lou A, Murtha T (1988) Environmental effects on glass fiber reinforced PPS stampable composites. J Mater Eng 10(2):109–116

Ma C-CM, Huang YH, Chang MJ (1991) Effect of jet fuel on the mechanical properties of PPSIC. F. and PeekIC. F. after impact loading. In: Tsai SW, Springer GS (eds) Proceedings of the eighth international conference on composite materials (lCCM/8), Composites design, manufacture, and application, Honolulu, pp 16-M-1–16-M-10

Mandell JF (1979) Origin of moisture effects on crack propagation in composites. Polym Eng Sci 19(5):353–358

Manocha LM, Bahl OP, Jain RK (1982) Performance of carbon fibre reinforced epoxy composites under different environments. In: Hayashi T, Kawata K, Umekawa S (eds) Proceedings of the fourth international conference on composite materials (ICCM/4), Progress in science and engineering of composites, vol 2, Tokyo, 25–28 October, pp 957–964

McKinnis CL (1978) Stress corrosion mechanisms in E-glass fiber. In: Bradt RC, Hasselman DPH, Lange FF (eds) Fracture mechanics of ceramics vol. 4, Plenum Press, New York London, pp 581–596

Menges G, Gitschner H-W (1980) Sorption behavior of glass-fibre reinforced composites and influence of diffusion media on deformation and failure behavior. In: Bunsell AR, Bathias C, Martrenchar A, Menkes D, Verchery G (eds) Proceedings of the third international conference on composite materials (ICCM/3), vol 1, Paris, pp 25–48

Metcalfe AG, Schmitz GK (1972) Mechanism of stress corrosion in E-glass filaments. Glass Technol 13(1):5–16

Metcalfe AG, Gulden ME, Schmitz GK (1971) Spontaneous cracking of glass filaments. Glass Technol 12(1):15–23

Morii T, Tanimoto T, Maekawa Z, Hamada H, Kiyosumi K (1991) Effect of surface treatment on degradation behavior of GFRP in hot water. In: Cardon AH, Verchery G (eds) Durability of polymer based composite systems for structural applications. Elsevier Applied Science, New York, pp 393–402

Morton J, Kellas S, Bishop S (1988) Damage characteristics in notched carbon fiber composites subjected to fatigue loading–environmental effects. J Compos Mater 22(7):657–673

Nakanishi Y, Shindo A (1982) Deterioration of CFRP and GFRP in salt water. In: Hayashi T, Kawata K, Unlekawa S (eds) Proceedings of the fourth international conference on composite materials (ICCM/4), vol 2, Tokyo. ISBS, Beaverton, pp 1009–1016

Norwood LS, Marchant A (1981) Recent developments in polyester matrices and reinforcements for marine applications, in particular in polyester/kevlar composites. In: Marshall IH (ed) Composite structures. Applied Science Publishers, New York, pp 158–181

O'Brien TK, Raju IS, Garber DP (1986) Residual thermal and moisture influences on the strain-energy release rate analysis of edge delamination. J Compos Technol Res 8:37–47

Ogi K, Takao Y (1998) Effect of moisture on stress-strain response and damage evolution in quasi-isotropic laminates. J Jpn Soc Compos Mater 24(1):20–29

Ogi K, Kim HS, Maruyama T, Takao Y (1999) The influence of hygrothermal conditions on the damage processes in quasi-isotropic carbon/epoxy laminates. Compos Sci Technol 59 (16):2375–2382

Perreux D, Suri C (1997) A study of the coupling between the phenomena of water absorption and damage in glass/epoxy composite pipes. Compos Sci Technol 57(9–10):1403–1413

Phillips DC, Scott JM, Buckley N (1978) The effects of moisture on the shear fatigue of fibre composites. In: Noton B, Signorelli R, Street K, Phillips L (eds) Proceedings of the 1978 international conference on composite materials (ICCM/2), Toronto, 16–20 April 1978, pp 1544–1559

Pomies F, Carlsson LA, Gillespie JW Jr (1995) Marine environmental effects on polymer matrix composites. In: Martin RH (ed) Composite materials: fatigue and fracture, vol 5, ASTM STP 1230. American Society for Testing and Materials, Philadelphia, pp 28–303

Price JN (1989) Stress corrosion cracking in glass reinforced composites. In: Roulin-Moloney AC (ed) Fractography and failure mechanisms of polymers and composites. Elsevier Applied Science, New York, pp 495–531

Ramirez F, Carlsson L, Acha B (2008) Evaluation of water degradation of vinylester and epoxy matrix composites by single fiber and composite tests. J Mater Sci 43(15):5230–5242

Roberts RC (1978) Design, strain and failure mechanisms of GRP in a chemical environment. Reinforced Plastics Congress, British Plastics Federation, London, Paper No. 19, pp 145–151

Russell AJ, Street KN (1985) Moisture and temperature effects on the mixed-mode delamination fracture of unidirectional graphite/epoxy. Delamination and debonding of materials, ASTM 537 STP 876, American Society for Testing and Materials, Philadelphia, pp 349–370

Saito H, Kimpara I (2007) Effect of water absorption on compressive strength after impact and post impact fatigue behavior of woven and knitted CFRP laminates. Key Eng Mater 334–335:517–520

Sandifer JP (1982) Effects of corrosive environments on graphite/epoxy composites. In: Hayashi T, Kawats K, Umekawa S (eds) Proceedings of the fourth international conference on composite materials (ICCM/4), vol 2: Progress in science and engineering of composites, Tokyo, 25–28 October, pp 979–986

Selvarathinam A, Weitsman Y (1998) Transverse cracking and delamination in cross-ply gr/ep composites under dry, saturated and immersed fatigue. Int J Fract 91(2):103–116

Sendeckyj GP (1990) Life-prediction for resin matrix composite materials. In: Reifsnider KL (ed) Fatigue of composite materials. Elsevier, Amsterdam, pp 431–483

Sheard PA, Jones FR (1986) The stress corrosion of glass fiber and epoxy composites in aqueous environments. In: Loo TT, Sun CT (eds) Proceedings of the international symposium on composite materials and structures, Beijing, 10–13 June 1986, pp 118–123

Shen CH, Springer GS (1981) Effects of moisture and temperature on the tensile strength of composite materials. In: Springer GS (ed) Environmental effects of composite materials. Technomic Publishing, Covina, pp 79–92

Siriruk A, Penumadu D, and Weitsman YJ (2010) Fatigue behavior of carbon fiber and vinyl ester sandwich facing material due to sea environment. In: Ninth international conference on sandwich structures (ICSS-9). California Institute of technology, Pasadena, 14–16 June 2010

Sloan FE, Seymour RJ (1992) The effect of seawater exposure on mode I interlaminar fracture and crack growth in graphite/epoxy. J Compos Mater 26(18):2655–2673

Smith LV, Weitsman YJ (1996) The immersed fatigue response of polymer composites. Int J Fract 82(1):31–42

Somiya S, Morishita T (1993a) Degradation phenomenon of fracture toughness of SMC immersed in acid condition. In: Chandra T, Dhingra AK (eds) Proceedings of the international conference on advanced composite materials (ICACM), Wollongong, Australia, pp 637–642

Somiya S, Morishita T (1993b) Study on the degradation phenomena of SMC in alkali condition by acoustic emission method. In: Miravete A (ed) Proceedings of the ninth international conference on composite materials (lCCM/9)", vol 5, Madrid. The Minerals, Metals, and Materials Society, Warrendale, pp 585–592

Springer GS, Sanders BA, Tung RW (1981) Environmental effects on glass fiber reinforced polyester and vinylester composites. In: Springer GS (ed) Environmental effects of composite materials. Technomic Publishing, Westport, pp 126–144

Strait L, Karasek M, Amateau M (1992) Effects of stacking sequence on the impact resistance of carbon fiber reinforced thermoplastic toughened epoxy laminates. J Compos Mater 26 (12):1725–1740

Sumsion HT (1976) Environmental effects on graphite-epoxy fatigue properties. J Spacecr Rockets 13(3):150–155

Talreja R (1981) Fatigue of composite materials: damage mechanisms and fatigue-life diagrams. Proc R Soc Lond A 378(1775):461–475

Talreja R (1990) Statistical considerations. In: Reifsnider KL (ed) Fatigue of composite materials. Elsevier, Amsterdam, pp 485–501

Thomas WF (1960) The strength and properties of glass fibers. Phys Chem Glasses 1(1):4–18

Van den Emde CAM, Van den Dolder A (1991) Comparison of environmental stress corrosion cracking in different glass fibre reinforced thermoset composites. In: Cardon AH, Verchery G (eds) Durability of polymer based composite systems for structural applications. Elsevier Applied Science, New York, pp 408–417

Verette RM (1975) Temperature/humidity effects on the strength of graphite/epoxy laminates, AIAA Paper No. 75–1011, AIAA 1975 Aircraft systems and technology meeting, Los Angeles, CA, 4–7 August 1975

Walker L, Hu XZ (2003) Mode I delamination behaviour of short fibre reinforced carbon fibre/ epoxy composites following environmental conditioning. Compos Sci Technol 63 (3–4):531–537

Weitsman YJ, Penumadu D, Siriruk A (2009) Immersed and dry fatigue behaviour of carbon fiber and vinyl ester sandwich facing materials. In: 17th International conference on composite materials (ICCM-17), Edinburgh, UK, 27–31 July 2009

White RJ, Phillips MG (1985) Environmental stress-rupture mechanisms in glass fiber/polyester laminates. In: Harrigan WC Jr, Strife J, Dhingra AK (eds) Proceedings of the fifth international conference on composite materials (lCCM/5), San Diego. Metallurgical Society of AIME, Warrendale, pp 1089–1099

Wyatt RC, Norwood LS, Phillips MG (1981) The stress-rupture behaviour of GRP laminates in aqueous environments. In: Marshall IH (ed) Composite structures. Applied Science Publisher, New York, pp 79–91

Yang B-X, Kasamori M, Yamamoto T (1992) The effect of water on the interlamina delamination growth of composite laminates. In: Sun CT, Loo TT (eds) Proceedings of the second international symposium on composite materials and structures, Beijing, 3–7 August 1992, pp 334–339

Zaffaroni G (1997) Two-dimensional moisture diffusion in hybrid composite components. In: High temperature and environmental effects on polymeric composites, Vol 2; Proceedings of the symposium, Norfolk, VA, 13 November 1995, pp 97–109

Zhao S, Gaedke M (1996) Moisture effects on mode II delamination behavior of carbon/epoxy composites. Adv Compos Mater 5:291–307

Zhurkov SN (1965) Kinetic concept of the strength of solids. Int J Fract Mech 1:311–323

Chapter 8
Sea Water Effects on Foam-Cored Sandwich Structures

8.1 General

Sandwich composite lay-ups are being utilized to an ever-increasing extent in naval structures. Typically, those lay-ups consist of closed-cell polymeric foams placed between fiber reinforced polymeric facings. The resulting sandwich components possess an exceedingly light weight, thereby increasing submersibles' buoyancy and, when employed in superstructural designs, may enhance a ship's stability by lowering its center of gravity.

However, like all other polymers and polymeric composites, sandwich lay-ups absorb water and are therefore susceptible to hydrothermal property degradations. The effects of sea water and humidity, in combination with temperature, are therefore of particular interest for naval structures since they are subjected to sustained exposures to these environments.

The evaluation of degradation in material properties is focused on the foam, the facings, and the foam/facing interface. That interface consists of a resin-rich region that bonds foam and facings together.

8.2 Ingress of Sea Water into Closed-Cell Polymeric Foams: Data and Observations

Earlier data on the uptake of sea water were collected for Divinycell H100 and H200 closed-cell PVC foams (Li and Weitsman 2004). Weight gains were recorded periodically for samples of thicknesses varying between 2 and 9 mm that were immersed in sea water at room temperature for more than 2 years. In addition, expansional strains were measured periodically and microscopic observations of fluid-induced damage were taken intermittently.

Y.J. Weitsman, *Fluid Effects in Polymers and Polymeric Composites*,
Mechanical Engineering Series, DOI 10.1007/978-1-4614-1059-1_8,

Fig. 8.1 Undamaged dry PVC foams (Li and Weitsman 2004)

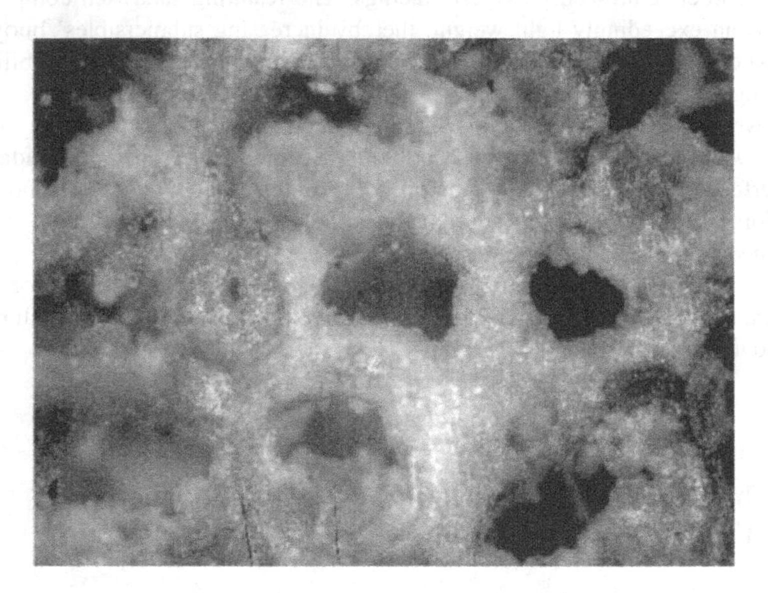

Fig. 8.2 Sea-water induced damage causing swelling of foam cell walls (Li and Weitsman 2004)

Typical observations of sea water-induced damage are shown in Figs. 8.1–8.3. Weight gain data are exhibited in Fig. 8.4a, b, with data scatter indicated by vertical bars.

Weight gain data at saturation varied most noticeably with the thickness of the foam samples, as shown in Fig. 8.5a, b for the H100 and H200 foams, respectively.

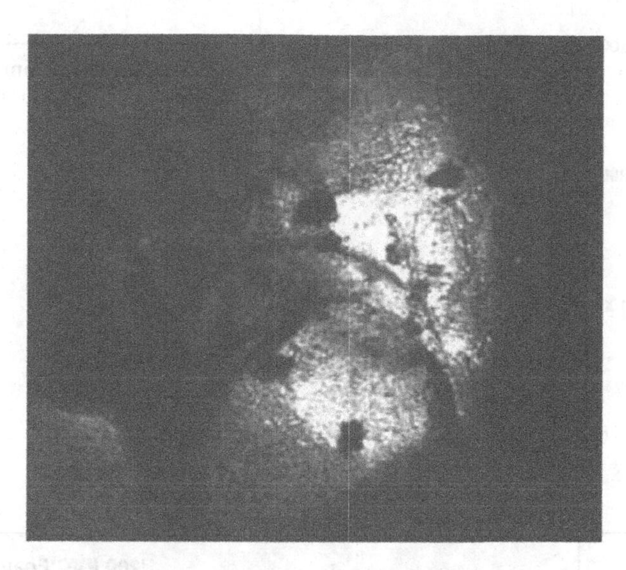

Fig. 8.3 A confocal microscope photograph shows pits forming at the bottom of a cavity (2.8 mm below the surface) (Li and Weitsman 2004)

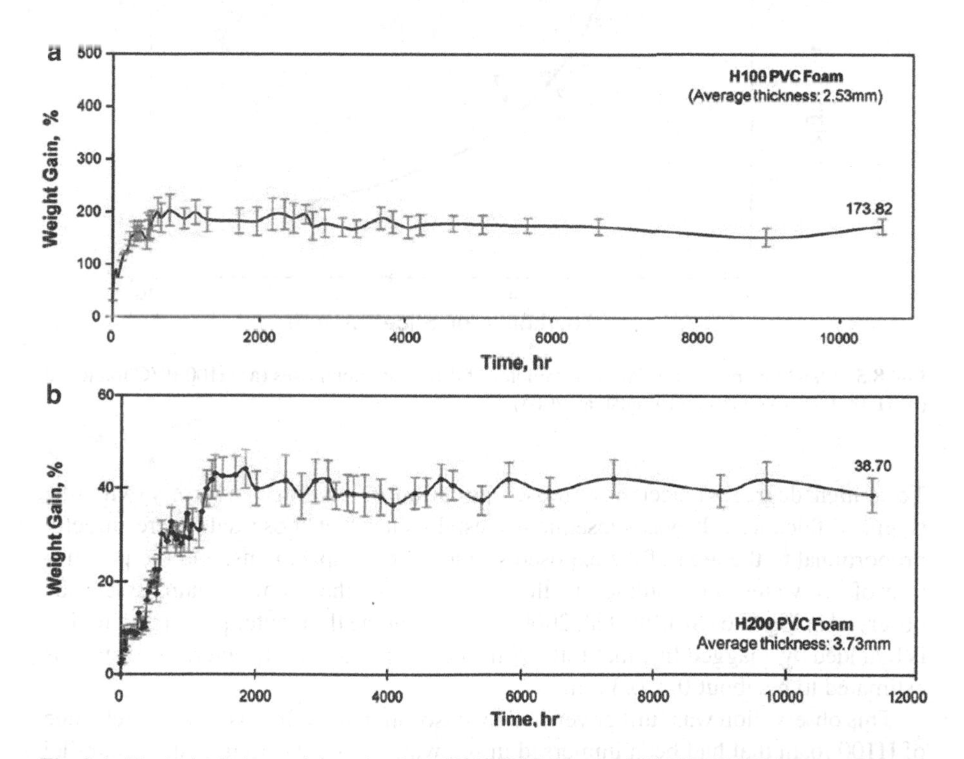

Fig. 8.4 Typical relative weight gain of (**a**) H100 PVC foam and (**b**) H200 PVC foam vs. time (Li and Weitsman 2004)

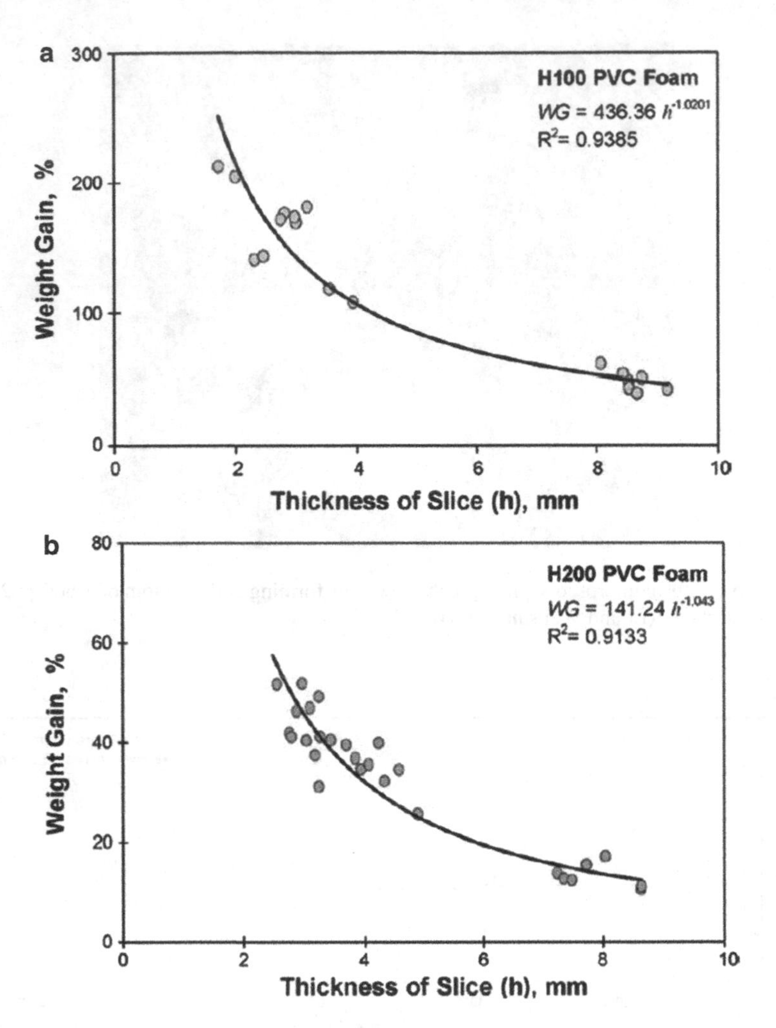

Fig. 8.5 Weight gain data for flat foam samples of different thicknesses (**a**) H100 PVC foam and (**b**) H200 PVC foam (Li and Weitsman 2004)

To a high degree of accuracy, those saturation levels vary inversely with the samples' thickness. It was subsequently established that those data were directly proportional to the area of the exposed surfaces of the specimens, and the penetration of sea water was confined to the outer cells of the foam domain, as can be observed in Fig. 8.6 (Siriruk et al. 2009a). Note that the thin, outer penetrated region is bounded by a jagged line that follows the cellular structure. Its average depth was estimated to be about 0.2–0.3 mm.

This observation was further verified by desorption data. In this case a thick cube of H100 foam that had been immersed in sea water was cut into thin slices parallel to its top and bottom faces that were subsequently placed within dry desiccating

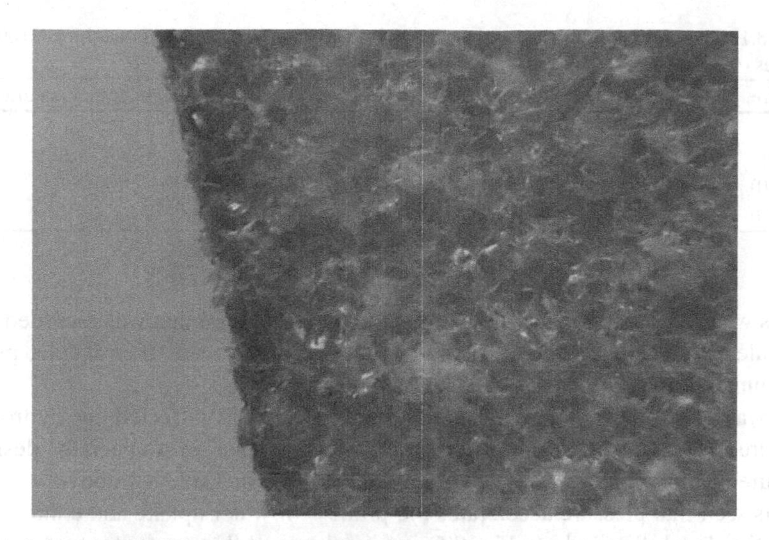

Fig. 8.6 A micrograph of an immersed foam, showing that the penetration of sea water remains confined near the exposed boundary after 3 years of immersion (Siriruk et al. 2009a)

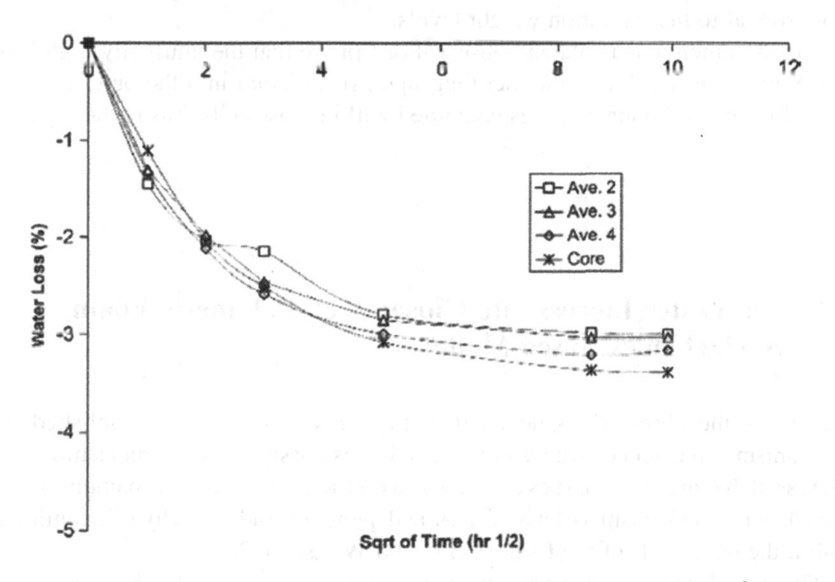

Fig. 8.7 Weight loss data for 2 mm thin slices cut from interior layers of a (25 mm)^3 H100 foam (*number* indicate slice location away from exposed surface). The weight loss from the outer layer was 96%

chambers. The resulting weight losses vs. time, exhibited in Fig. 8.7 (Siriruk 2009), demonstrate that the dominant portion of sea water was indeed contained in the outermost slice, where weight loss was about 96%. In addition, this figure shows that smaller amounts of water were contained within the interior layers. The ingress

Table 8.1 The effect of hydrostatic pressure on the ingress of sea water in 4 mm thick H100 foam samples

Pressure, p in kPa (psi)	Time to saturation (h)	Weight gain at saturation %
0	1,400	95
70 (10)	170	152
140 (20)	70	149
280 (40)	<60	147

of this water was attributed to the permeability of the foam that was recorded to be equivalent to a flow rate of 1.2 L/min at a 5-m water head[1] (Penumadu, private communication).

It was also noted that sea water ingress is strongly affected by hydrostatic pressure. The immersion of H100 4-mm thick samples in a specially designed pressure chamber yielded the weight gain data listed in Table 8.1 above.

It is seen that pressure accelerates the process of water uptake and enhances the saturation level. Similarly to Fig. 8.6, it was observed that sea water was confined within a thin layer of foam cells adjacent to the exposed boundary. Both observations and elementary considerations suggest that the depth of that layer is proportional to the saturation weight levels.

The aforementioned data and observations prove that the unusually high levels of weight gains are due to the fact that, upon penetration into the outer cells, sea water fills the entire inner spaces contained within those cells thus replacing air by water.

8.3 Sea Water Ingress into Closed-Cell Polymeric Foam. A Mechanics-Based Model

Guided by the observations depicted in Figs. 8.4 and 8.6, that established both mechanism and extent of sea water ingress, it was possible to construct a model that addressed the three basic issues associated with the ingress process, namely (i) what is its driving mechanism, (ii) why is it time dependent, and (iii) why is it confined to within the outer cells of the foam (Ionita and Weitsman 2007).

The model generated several sets of random patterns of cellular domains, with cells of varying sizes that mirrored those of the actual foam. However, due to

[1] Diffusion across the thickness of the foam layers, which could proceed along the cellular walls, is judged to have a miniscule effect on weight gain. This is due to the fact that for typical polymers it would require years if not decades to attain diffusional equilibrium for the thicknesses involved herein.

limitations on computational capacity, the above patterns were two dimensional – in contrast with the three dimensionalities of the actual foam cells.[2]

In view of the confined extent of water penetration, which is certainly true for early exposure times, the associated water-induced swelling is constrained by the presence of a larger and stiffer interior dry domain. This constraint against free swelling induces compressive stresses within the walls of the penetrated cells that may exceed their bucking levels. Note that this proposition is supported by the recorded effects of hydrostatic pressure listed in Table 8.1.

These buckling amplitudes would vary from wall to wall and cell to cell due to the random variations in their dimensions. Once buckling occurs, the wall is breached and sea water gushes into the adjacent cell. Such breached walls can indeed be seen in Fig. 8.2. This is postulated to be the sequential ingress mechanism of sea water into the closed-cell polymeric foam.

The time dependence of the ingress process was attributed to the water-enhanced viscoelastic response of the cellular walls. Accordingly, even if any wall dimensions are sufficiently small to withstand buckling at the instant of the exposure to sea water, failure may still occur as the time-dependent relaxation modulus $E(t)$ decreases to a critical level of $E_c = E(t_c) < E(0)$ at time $t = t_c$ subsequent to its initial exposure to sea water.

The time dependence of the ingress process could therefore be correlated with the time-delayed buckling process of several cellular walls.

Note that for walls just several microns thick, diffusion would attain its equilibrium level within about 1 h. This timespan may be considered as instantaneous relative to the thousands of hours required to terminate the ingress process. Time delay is thus due to viscoelasticity and much less so due to diffusion.

Finally, water ingress terminates at a boundary where all cellular walls are sufficiently short to withstand buckling even at $E(\infty)$.

The computational scheme had to accommodate the presence of time-varying saturated domains (i.e., time varying boundary conditions) as well as keeping track of all the disparate initial exposure times t_i, that varied from wall to wall, in order to monitor values of $E(t - t_i)$ for each cell – with its own individual dimensions – to check the presence or absence of failure by buckling. Note that an entirely revised stress analysis was necessary for every new configuration of the unbuckled, dry, or saturated domain.

Computational results are shown in Figs. 8.8 and 8.9. Figure 8.8, which exhibit the sequential penetration of sea water into a specific randomly chosen configuration of a two-dimensional cellular structure at various times, while Fig. 8.9 shows presumed weight gain levels vs. time at five distinct randomly chosen cellular configurations. Note that similarly to actual multisample data, the latter figure reflects data scatter.

[2] Despite this serious limitation, the model still explains the essential features of the ingress phenomenon.

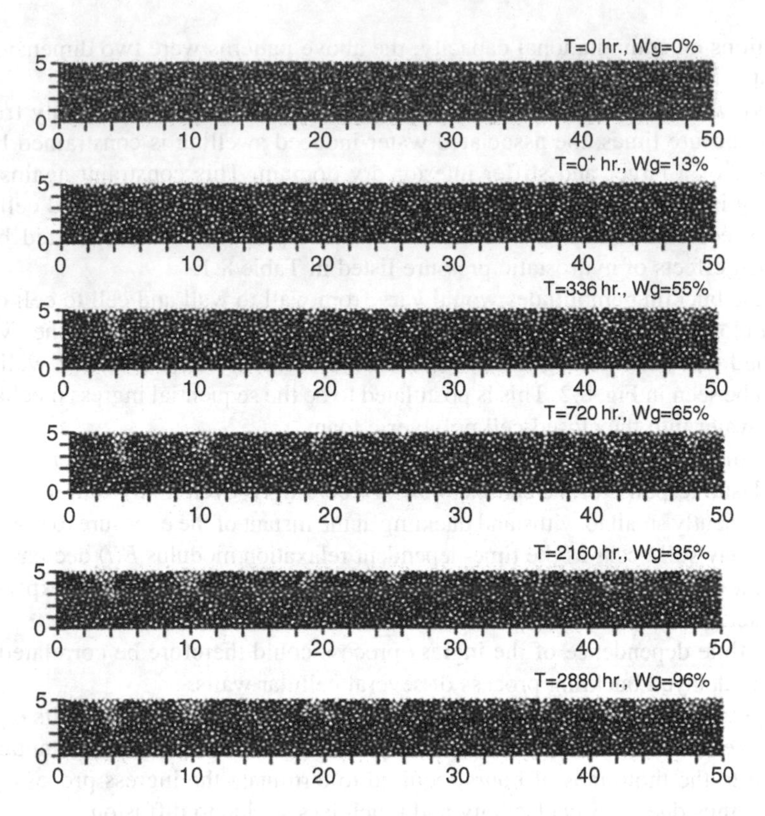

Fig. 8.8 Predicted profiles of water ingress into cellular foam after several exposure times

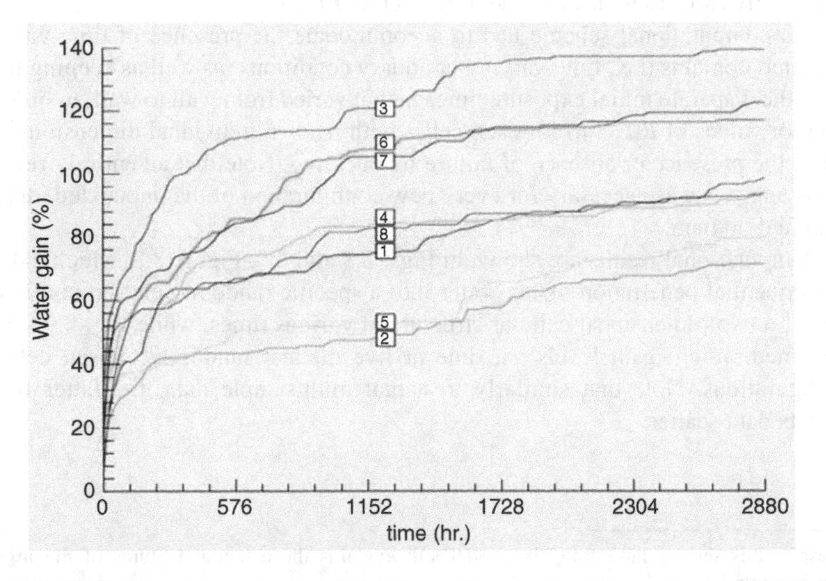

Fig. 8.9 Predicted weight gains in cellular foams for several random selections of Voronoi cellular configurations

8.4 Foam Testing and Material Data

The mechanical testing of foams requires several adaptations due to their low strengths and stiffnesses. These are mostly concerned with the method of load introduction and the selection of appropriately sensitive load cells.

Typical dumbbell-shaped test specimens, such as that shown in Fig. 8.10 (Siriruk et al. 2009a), were chosen for the performance of tensile and torsional tests with loads introduced through friction over the top and bottom circular portions.

Two techniques, "direct" shear and torsion, were used to test foam samples in shear as shown in Figs. 8.11 and 8.12. Tensile and shear stress strain data for H100 foam are shown in Figs. 8.13 and 8.14, respectively. The initial slopes in those figures correspond to $E = 60$ MPa and $G = 27$ MPa. Note that those values are averaged over a scatter of about $\pm 6\%$. It was also noted that stiffnesses increase monotonically toward the center of a 25 mm thick foam slab.

An estimate for the stiffnesses within the saturated outer region can be obtained from torsional data on coupons that were preimmersed to saturation. A sketch of a cross section of a test sample with dry inner core and a saturated boundary zone is shown in Fig. 8.15 and a sample torsional data for a coupon that was tested in the dry state and subsequently at equilibrium saturation weight is shown in Fig. 8.15. It follows from the torque/rotation plots in Fig. 8.16 that for that particular sample $\bar{G}_w \sim 0.84\, G$, where $\bar{G}_w$ denotes the shear modulus of the immersed sample, averaged over its cross section.

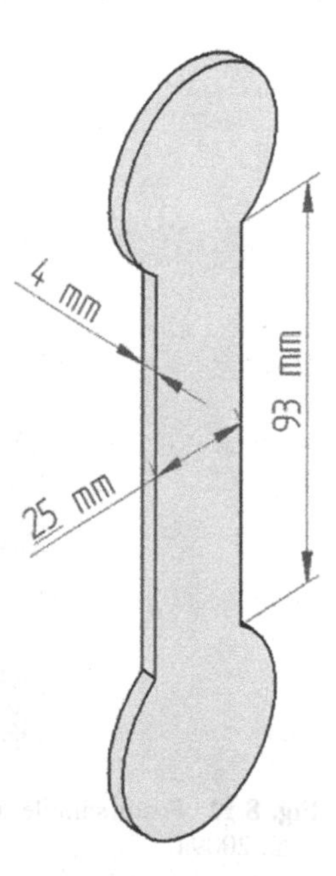

Fig. 8.10 Foam specimen shape and geometry

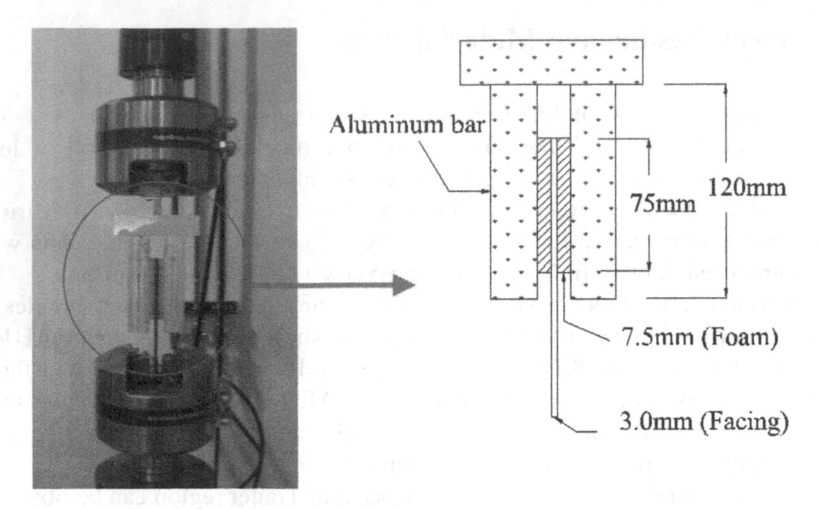

Fig. 8.11 Shear response experimental set up (Siriruk et al. 2009a)

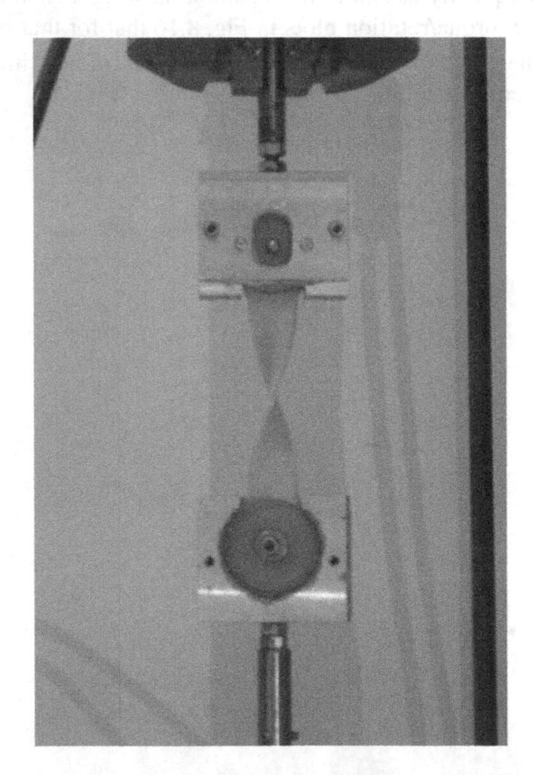

Fig. 8.12 Foam sample undergoing large torsional rotation (rotary actuator at the *top*) (Siriruk et al. 2009a)

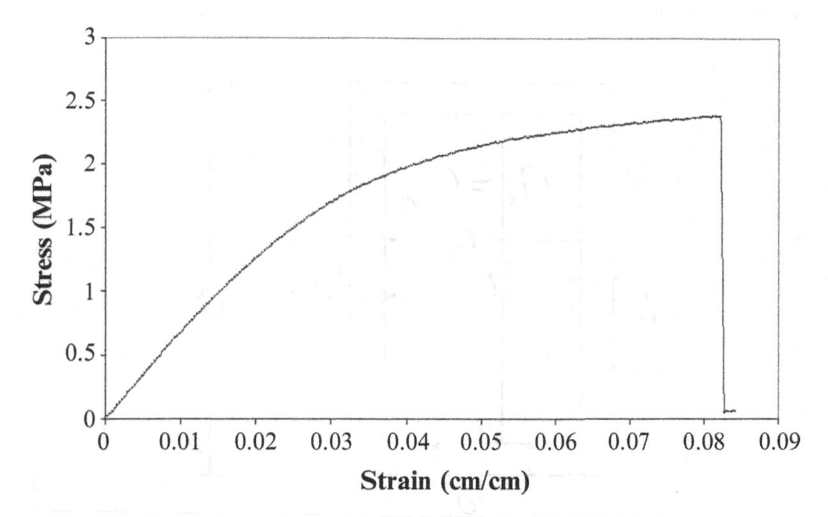

Fig. 8.13 Tensile stress-strain data ($E_{\text{initial}} = 60$ MPa) (Siriruk et al. 2009a)

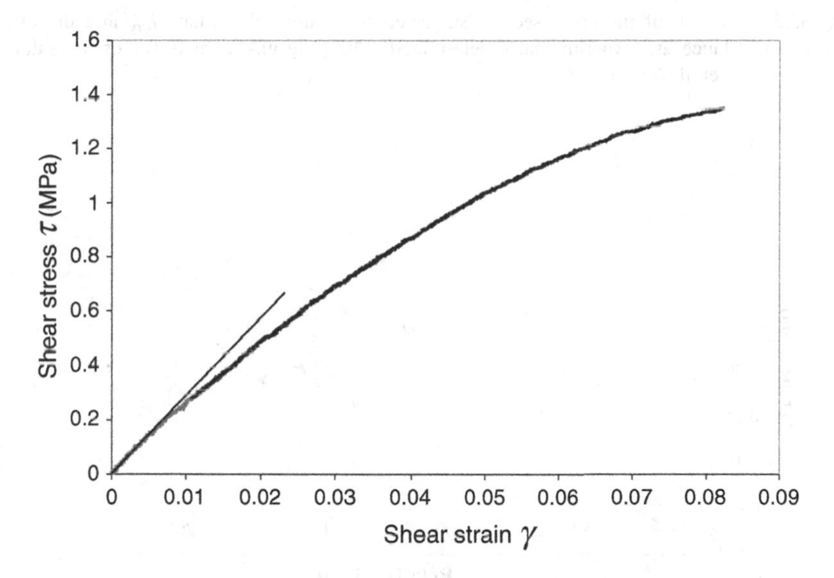

Fig. 8.14 Stress-strain data in shear $G_{\text{initial}} = 26.5$ MPa (Siriruk et al. 2009a)

It is possible to assess the value of G_w within the saturated outer layer alone by means of a linear elastic torsional analysis for a rectangular cross section of height H and width W that consists of an outer saturated "frame" of thickness δ and an inner dry core of dimensions $H - 2\delta$ and $W - 2\delta$, respectively (Timoshenko and Goodier 1970). For the specific values of H, W, δ, G and G_w corresponding to Fig. 8.16, a straightforward calculation yields $G_w \sim 0.3 - 0.4\ G$. The above estimate utilized

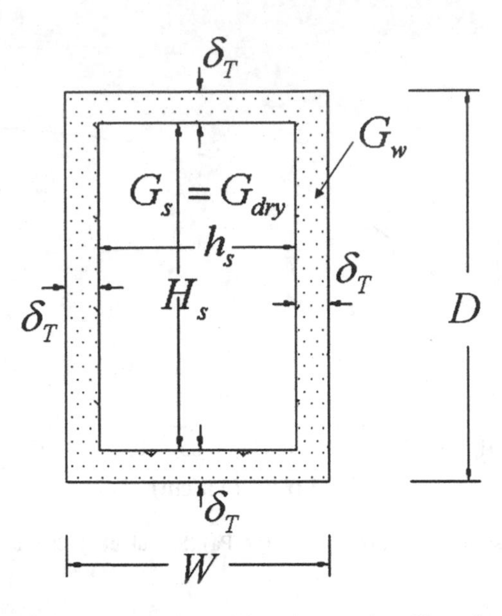

Fig. 8.15 A sketch of the cross section subjected to torsion. Note that G_{dry} and the overall torsional resistance are experimentally determined. The only unknown is G_w of the saturated region (Siriruk et al. 2009b)

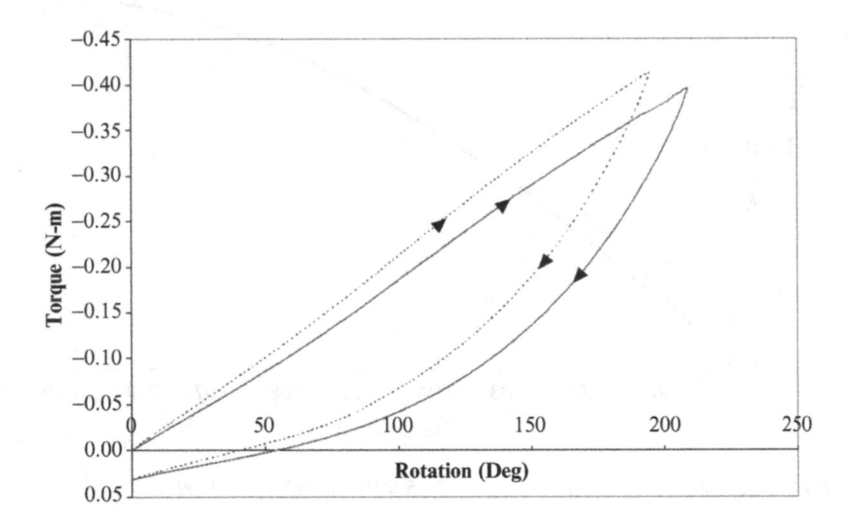

Fig. 8.16 Torque vs. rotation angle for dry (*broken line*) and saturated (*continuous line*) samples (Siriruk et al. 2009b)

the average value of δ while in reality the separation between the inner dry core and the surrounding saturated region occurs over a jagged interface in three dimensions. Since this computation applies to a presumed straight interface, it may be indicative only.

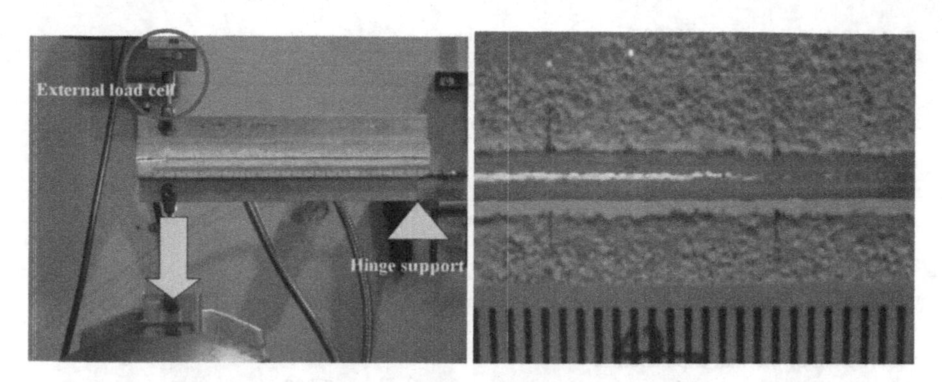

Fig. 8.17 Experimental setup (*left*) and observed crack propagation (*right*) for mode I fracture of foam (Siriruk 2009)

It should be recognized that due to the large data scatter, the values of $\bar{G}_w/G$ range between 0.84 and 0.95. Consequently, the estimated values of G_w/G may range between 0.3 and 0.9.

Hydral-induced expansion was inferred from shrinkage data collected upon drying from the saturated state.

Following an established procedure (Anderson 2004), mode I fracture toughness G_{IC} was evaluated from load/unloaded data under the sequential intermittent growth of a central crack along the midplane of a symmetrically edge-loaded H100 foam sample shown in Fig. 8.17. The sample, of dimensions 25 × 25 × 200 mm, was precracked at its loaded end and symmetrically notched along its entire midplane to channel the fracture direction.

To obtain the "wet" G_{IC}, it was necessary to reimmerse the sample in simulated sea water for at least 2 weeks after each intermittent crack growth in order to ascertain the existence of a saturated crack tip region prior to reloading. The experimental results are summarized in Table 8.2.

8.5 Facings Properties[3]

Two types of polymeric facing materials that are utilized in naval sandwich structures were considered.

The first type consisted of two glass/vinylester facing lay-ups of $[0/\pm45_3]$, with $V_f = 0.53$ and $V_f = 0.56$, in combination with the H100 and H200 foam cores, respectively. At saturation, those facings absorbed 0.2% of sea water. No mechanical properties were recorded.

[3] Although the information contained herein falls within the scope of Chap. 4 it is presented in this section as part of the inclusive discussion of sandwich layups.

Table 8.2 Summary of H100 foam core properties

Property	Foam	Dimension
Longitudinal modulus, E (Dry)	60	MPa
Longitudinal modulus, E (Wet) due to sea water[a]	22	MPa
Shear modulus, G (Dry)	25	MPa
Shear modulus, G (Wet) due to sea water[a]	7.3–23	MPa
Shear modulus, G (Wet) due to combined sea water and low temperature	22	MPa
Shear modulus, G (Wet) due to hydrostatic pressure[a]	23	MPa
Shear modulus, G (Wet) due to hydrostatic pressure and low temperature[a]	22	MPa
Coefficient of thermal expansion, α	70	$\mu\varepsilon$/°C
Moisture expansional strain at saturation, ε_H	2,200	$\mu\varepsilon$ at saturation
Poisson's ratio, ν (Dry)	0.13	
Poisson's ratio, ν (Wet)	0.50[a]	
Critical Mode I energy release rate G_{IC} dry, average (standard deviation)	835 (260)	N/m^2
Critical Mode I energy release rate G_{IC} wet, average (standard deviation)	895 (115)	N/m^2

[a] Implied from data by means of analysis

Table 8.3 Summary of carbon reinforced vinyl ester facing data

Property		Dimension
Longitudinal modulus E of $[0/90]_{2s}$	80	GPa
Longitudinal modulus E of $[\pm 45]_{2s}$	15	GPa
Longitudinal modulus E of $[0/90]_{2s}$ due to sea water	80	GPa
Longitudinal modulus E of $[\pm 45]_{2s}$ due to sea, tap, and distilled water	15	GPa
Longitudinal modulus E of $[\pm 45]_{2s}$ at -15°C	16	GPa
Coefficient of thermal expansion, α	11.5	$\mu\varepsilon$/°C
Moisture expansional strain at saturation, ε_H	450[a]	$\mu\varepsilon$ at 0.5% saturation weight gain

[a] This weight gain value was measured elsewhere to be 0.6%

The second facing material consisted of a stitched cross ply carbon fiber lay-up embedded within vinylester resin, which was tested in both [0°/90°] and $[\pm 45]_{2s}$ orientations.

The data are summarized in Table 8.3.

8.6 Delamination of the Core/Facing Interface

The delamination toughness at the core/facing interface was measured analogously to the technique listed at the end of Sect. 8.3.

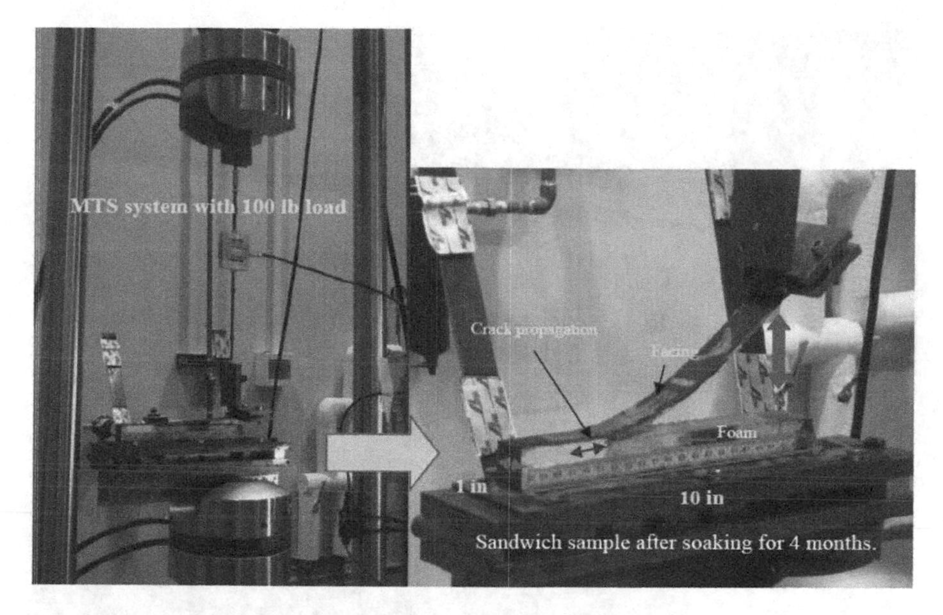

Fig. 8.18 Delamination testing setup using 0.44 kN load cell (Siriruk et al. 2009a)

A custom made set up, motivated by the tilted sandwich debond method (Li and Carlsson 1999), as shown in Fig. 8.18 (Li and Weitsman 2004; Siriruk et al. 2009a, b). In this case, the bottom facing of a 254-mm long sandwich sample was glued firmly to a metal block, while one end of the top facing was attached to loading machine by a hinged metallic device. A precrack of prescribed length was precut into the interface prior to testing and G_{IC} was evaluated from in the same manner and noted in Sect. 8.4.

Note that for the wet case, it was necessary to reimmerse the sample for at least 2 weeks past every intermittent crack growth to ascertain the presence of a saturated crack tip zone. A photo of a dry and wet debonding test is shown in Fig. 8.19.

In view of the small depth of the wet tip zone, most of the crack growth occurs within a dry region. Consequently, the immersed fracture data relate crack initiation energy.

Comparative critical values of dry and immersed debonding fracture energies (per unit area) are listed below:

For glass/vinylester facings (Li and Weitsman 2004)

$$\text{H100} \quad G_{\mathrm{c}}^{\mathrm{dry}} = 598(\mathrm{J/m^2}), \quad G_{\mathrm{c}}^{\mathrm{immersed}} = 786(\mathrm{J/m^2}),$$

$$\text{H200} \quad G_{\mathrm{c}}^{\mathrm{dry}} = 1,128(\mathrm{J/m^2}), \quad G_{\mathrm{c}}^{\mathrm{immersed}} = 1,218(\mathrm{J/m^2}).$$

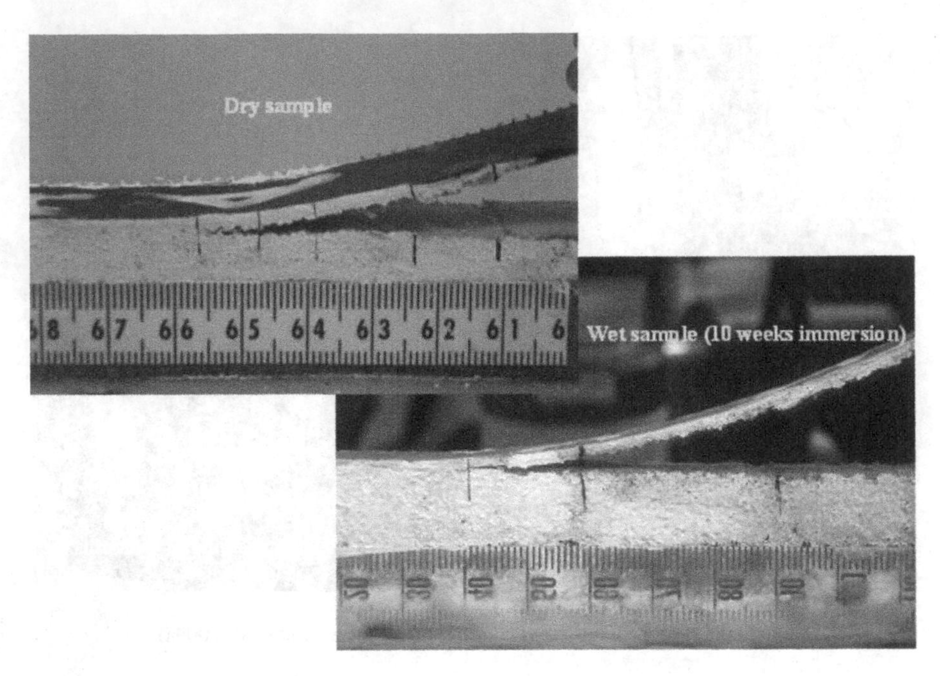

Fig. 8.19 A detailed photograph of a dry and "wet" delamination crack

Table 8.4 Experimental values of wet and dry interlaminar fracture toughness G_c for H100/carbon/vinyl ester facing

	Lower & upper limit G_c (N/m)	Representative range (>50%) (N/m)
Dry	560–955	780–890
Wet	440–625	522–588
Reduction (%)	21–35	33–34

The analogous data for the carbon vinyl/ester facing with H100 foam are listed in Table 8.4 (Siriruk et al. 2009b).

The contrasting trends among the above facings may be due to the fact that the glass fibers are susceptible to fluid effects. Differences may also be attributed to variations in the degree of cure, as well as to the observation that with the glass fiber facing the modulus of the H100 foam was recorded at $E = 95$ MPa.[4]

In both circumstances, it was observed that for the immersed conditions, the debonding cracks tended to progress closer to the interface, while in the dry case they advanced within the foam further away from it. A detailed fracture mechanics analysis (Li and Weitsman 2004) has shown that G_c consisted of both G_I and G_{II} fracture modes, and while K_{II} was positive in the dry case it was negative in the

[4] The sandwich samples containing carbon and glass fiber facings were provided by different suppliers.

immersed circumstance. This explains the observation of the trend for the growth of the debonding cracks, namely toward and away from the interface.

References

Anderson TL (2004) Fracture mechanics: fundamentals and applications, 3rd edn. CRC Press, Boca Raton

Ionita A, Weitsman Y (2007) A model for fluid ingress in closed cell polymeric foams. Mech Mater 39(5):434–444

Li X, Carlsson LA (1999) The Tilted Sandwich Debond (TSD) specimen for face/core interface fracture characterization. J Sandwich Struct Mater 1(1):60–75

Li X, Weitsman YJ (2004) Sea-water effects on foam-cored composite sandwich lay-ups. Compos B Eng 35(6–8):451–459

Siriruk A (2009) The mechanical characterization of polymeric sandwich materials for marine applications. Ph. D. Dissertation, University of Tennessee, Knoxville

Siriruk A, Weitsman YJ, Penumadu D (2009a) Polymeric foams and sandwich composites: material properties, environmental effects, and shear-lag modeling. Compos Sci Technol 69 (6):814–820

Siriruk A, Penumadu D, Weitsman YJ (2009b) Effect of sea environment on interfacial delamination behavior of polymeric sandwich structures. Compos Sci Technol 69(6):821–828

Timoshenko S, Goodier J (1970) Theory of elasticity. McGraw-Hill, New York

Chapter 9
Special Issues

9.1 The Reverse Thermal Effect

9.1.1 Background and Existing Interpretations

Useful insight into the mechanism of fluid ingress into polymers and polymeric composites can be gained by noting and interpreting the so-called "reverse thermal effect." The earliest discovery of this phenomenon was made independently and interpreted differently (Apicella et al. 1979; Adamson 1980).

Additional interpretations were offered more recently (Weitsman 1990; Suh et al. 2001), and a possible hypothesis by the author is presented as well.

The phenomenon concerns a somewhat bewildering result stemming from weight gain data recorded during temperature drop.

For several materials, it was noted that water uptake increased with temperature level. This rather typical observation is shown by the early stage of the middle curve in Fig. 9.1.

One would thus expect that upon lowering the temperature during an ongoing absorption process the corresponding weight gain level would decrease as well. However, the opposite takes place, as several repeatable data show that temperature drops are accompanied by an increase in fluid uptake. Note that, as may be expected, a rise in temperature is accompanied by an increase in fluid uptake. In summary, no matter if one raises or lowers the temperature during a progressing absorption process, the result will always be an increase in the fluid content. This is demonstrated in Fig. 9.2.

In Adamson's view (Adamson 1980), the free volume within a polymeric material increases as temperature drops, because the thermal vibrations of the polymeric molecules decline, thereby reducing the occupied space. This increase in free volume thus explains the increase in weight gain. In addition, Adamson suggested that at any given temperature, water molecules are divided between bound and

Y.J. Weitsman, *Fluid Effects in Polymers and Polymeric Composites*,
Mechanical Engineering Series, DOI 10.1007/978-1-4614-1059-1_9,

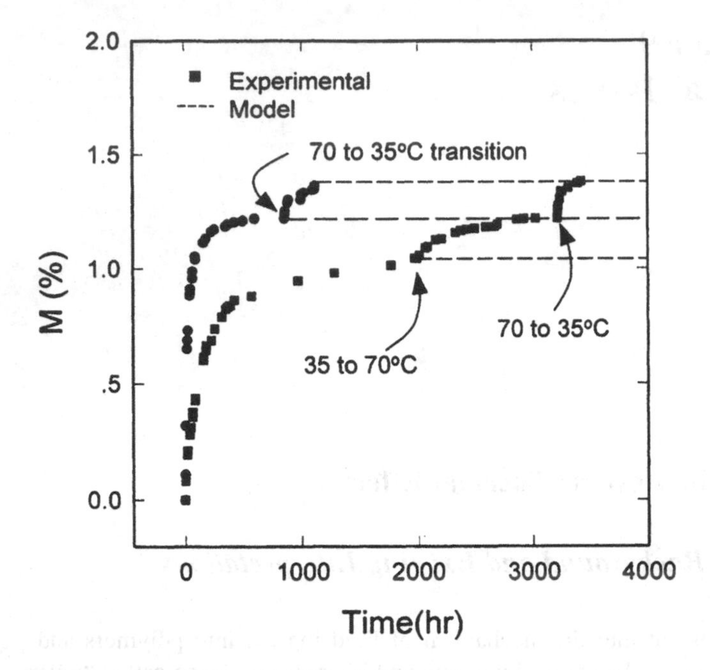

Fig. 9.1 Comparison of 70–35°C and 35–70–30°C hygrothermal cycles
Credit: Suh D, Ku M, Nam J, Kim B, Yoon S. "Equilibrium water uptake of epoxy/carbon fiber composites in hygrothermal environmental conditions." 35(3):264–278, copyright 2001 by *Journal of Composite Materials*. Reprinted by Permission of SAGE

mobile components, with the former causing swelling and the latter occupying the free volume. The rate of increase in the amount of bound molecules slows down as they are directed into the higher density micelles that remain in the polymer after the initial stage of absorption. As temperature drops, the bound water decreases the packing efficiency of the polymeric chains, causing them to expand. This expansion, which is compensated by thermal shrinkage, explains the observation that only negligible volumetric changes are recorded during the temperature drop.

Apicella et al. (1979) also assume that water within the polymer is divided between bound and free components that are associated with the polymeric chains and microcavities, respectively. They formulate a model for the growth of microcavities that enables them to explain the reverse thermal effect.

Yet another explanation of the reverse thermal effect was offered by Suh et al. (2001) for a carbon/epoxy composite. They also consider that absorbed water consists of free and bound components, M_F and M_B, respectively. Similarly to previous considerations (Vieth and Sladek 1965; Vieth et al. 1976), the authors relate the bound water by means of Henry's law and the free water by means of Langmuir's law. They assume that the sorption of bound water is a reversible process and consider the uptake of free water to be irreversible – thereby not being squeezed out upon changes in hygrothermal conditions. It thus follows that

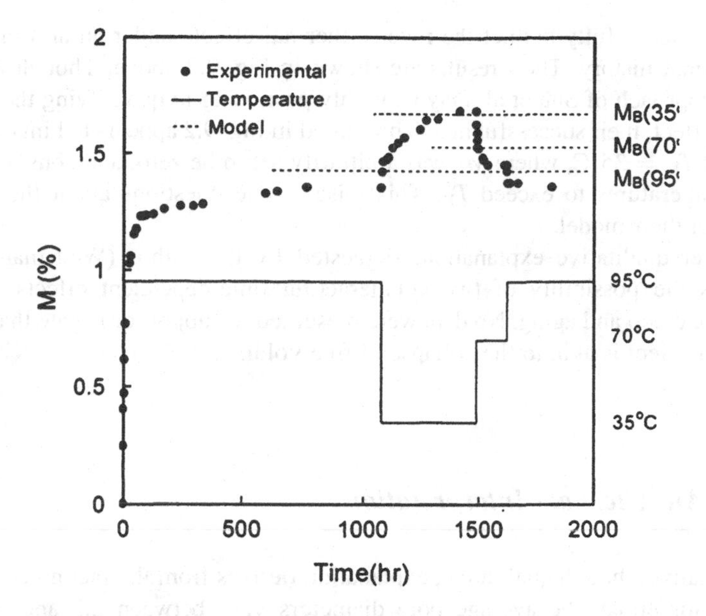

Fig. 9.2 Model comparison of equilibrium water uptake in 95–35–70–95°C hygrothermal condition
Credit: Suh D, Ku M, NamJ, Kim B, Yoon S. "Equilibrium water uptake of epoxy/carbon fiber composites in hygrothermal environmental conditions." 35(3):264–278, copyright 2001 by *Journal of Composite Materials*. Reprinted by Permission of SAGE

the amount of free water can never decrease. In agreement with Adamson (1980), they argue that the free volume diminishes with temperature. Their model is based on the aforementioned crucial assumptions.

Consequently, the amount of bound water is determined by the latest exposed temperature T_L and that of the free water by the highest exposed temperature T_H. Thus

$$M = M_B(T_L) + M_F(T_H) \tag{9.1}$$

which is further detailed as

$$M - M_{BO}\exp(\Delta H/RT_L) + M_{FO}\exp[-G_r/R(T_H - T_R)]. \tag{9.2}$$

In (9.2), ΔH denotes heat of sorption, G_r is the free energy of free volume sites (excluding elastic expansional energy of those sites), and T_R is the reference temperature at which there is no free water, i.e., $M_F(G_R) = 0$.

Considering $T_R = 35°C$, Suh et al. matched M_B and M_F with data over the range of $35°C \leq T \leq 95°C$, with individual temperature drops from 60, 70, 85, and 95°C down to 35°C for which M_F increase with T_H and M_B is adjusted to decrease (but less significantly) with T_L. Tabulating their results, they employed them as base values to successfully predict the reverse thermal effects under an arbitrary step-wise thermal history. Their results are shown in Fig. 9.2 above. Though it seems that the approach of Suh et al. may be highly promising in quantifying the reverse thermal effect, their successful match presented in Fig. 9.2 appears to hinge on their choice of $T_R = 35°C$, where M_F was arbitrarily set to be zero, and considering all other temperatures to exceed T_R. This raises some questions about the general validity of their model.

Another qualitative explanation, suggested by the author (Weitsman 1990), considers the possibility of two counteracting time-dependent effects, namely polymeric creep and aging. No data were presented to support or negate that thesis. The latter effect is akin to the collapse of free volume.

9.1.2 An Alternate Interpretation

An alternative, though qualitative, explanation derives from the fact noted in Sect. 3.1 that for epoxy the average pore diameters vary between 2.7 and 5 Å and assumes that, as in the case of polypropelene, actual diameters are distributed in some form about those averages. It follows that water molecules, with estimated diameters between 1.5 and 3 Å, cannot occupy the entire free volume contained within epoxies, either because some micropores are too small or because some larger micropores are connected by "microtubes" that are too narrow for the passage of water molecules. Those assumptions lead to the hypothesis that *a portion of the free volume remains inaccessible to water molecules* (Weitsman 2009, Unpublished work).

The other hypothesis is that *the effect of an abrupt temperature drop of environmentally exposed polymeric material is equivalent to the application of a sudden external isotropic pressure*. The latter hypothesis derives from the fact that both above factors induce isotropic compression, as correlated by $p = -E\alpha\Delta T$.

While the applied pressure causes an overall shrinkage of the molecular epoxy structure, it also activates motion of the water molecules. Since the shapes and sizes of the water-occupied micropores vary statistically, as do the configurations of the overall polymeric framework and the "microtube" structures, the applied isotropic external stress could well translate into many inhomogeneous centers of pressure of disparate amplitudes acting on highly anisotropic neighborhoods on the micro-level.

The localized centers of pressure would therefore squeeze water molecules along preferred directions into yet unoccupied and inaccessible portions of the free volume.

Although some molecules could well be squeezed out of the polymer from regions adjacent to its outer boundary, this effect is likely to be confined to a boundary layer only.

Furthermore, according the (5.36) and (5.42), the chemical potential $\tilde{\mu}$ increases with external pressure and thereby enhances the water absorption process. This may well be equivalent to the effect of a temperature drop.

The current proposition seems to be supported by weight gain data collected under two distinct amplitudes of temperature drop ΔT (Suh et al. 2001). Those data suggest that weight gains depend essentially on ΔT and just marginally, if at all, on T itself.

The observation that in several circumstances, the saturation level M_∞ increases with T may well be due to the overall expansion of the polymeric chains that enhances the accessibility of new micropores to water molecules. The above observation is not contradicted by any fundamental theoretical considerations and could be distinct from the mechanisms triggered by cooling.

9.2 Effects of Polymer Aging on Fluid Uptake

As noted earlier in Sect. 2.2.1, polymers in their glassy state are not in thermodynamic equilibrium and their aging is caused by the spontaneous, time-dependent, reduction of free volume. It is thus to be expected that fluid uptake, which is strongly related to free volume, should depend on the "shelf time" t_e prior to initial exposure and diminish as t_e increases.

It was possible to assess the effects of aging of two glass-fiber and one carbon fiber-reinforced composites for $\Delta t_e = 32$ weeks by recording their weight gain data at $t_e = t_0$ and $t_e = t_0 + \Delta t_e$ (Gao and Weitsman 1998). The value of t_0, i.e., the time lapse since material production, was estimated to be 2 weeks. Results are shown in Figs. 9.3–9.5.

These figures demonstrate that time aging plays a significant role on fluid uptake for the E-glass/NCT301 composite, which is distinct from "moisture aging" for all composites. Results similar to those for the E-glass/SP1003 composite were obtained for AS4/3501-6 graphite/epoxy. While moisture aging results in an increasing fluid uptake in most cases, timeaging is expected to decrease that amount.

9.3 Damage Susceptibility of Glass Fibers

Glass fibers come in several forms and possess different stiffnesses and strengths. These depend on the relative amounts of material components that they contain, with the largest portion being silicon oxide (SiO_2) that varies between 50 and 70% (Charles 1958a, b; Metcalfe et al. 1971; Metcalfe and Schmitz 1972). These fibers contain edge flaws of different sizes whose distribution may be correlated with tensile static fatigue data by means of the Griffith fracture criterion (Charles 1958a, b).

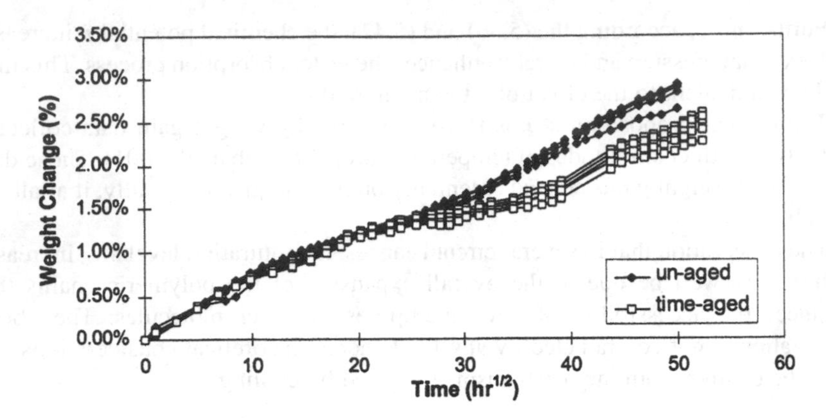

Fig. 9.3 Comparison of weight gain data for time-aged and un-aged E-glass/NCT301 coupons immersed in simulated sea water at 50°C. The time-aged coupons were immersed after a hold period of 32 weeks

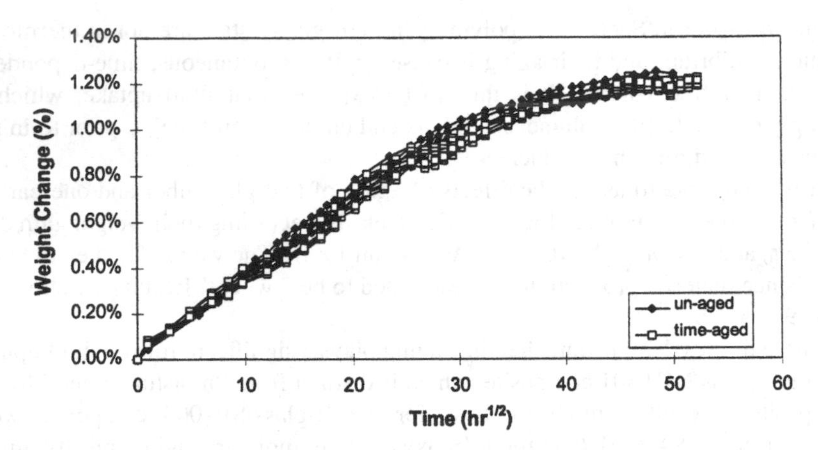

Fig. 9.4 Comparison of weight gain data for time-aged and un-aged AS4/3501-6 coupons immersed in simulated sea water at 50°C. The time-aged coupons were immersed after a hold period of 32 weeks

Exposure to fluids activates chemical interactions with some specific components of the glass fiber, varying with the particular chemistries of the fluid, (e.g., Sheard and Jones 1986). In many circumstances, that fluid ingress follows a sharp reaction front, e.g., lime glass rods exposed to saturated water vapor (Charles 1958a, b; Regester 1969). Furthermore, fluid penetration into the initial flaws enhances and accelerates their growth, thus hastening fiber failure. The corrosive effects of fluids can be detected by identifying the corrosion products that are extracted from the glass and by observing the large volumetric expansion of the corroded region (Charles 1958a, b) as well by the presence of microscopic grooves etched into the fiber boundary (Prian and Barkatt 1999).

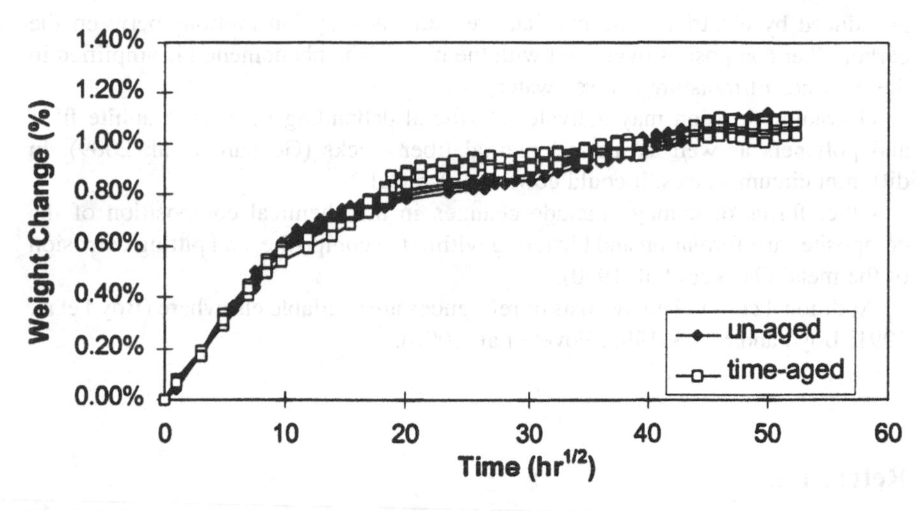

Fig. 9.5 Comparison of weight gain data for time-aged and un-aged E-glass/SP1003 coupons immersed in simulated sea water at 50°C. The time-aged coupons were immersed after a hold period of 32 weeks

In the case of glass re-inforced polymeric composites, it was occasionally observed that exposure to various fluids resulted in the formation of regularly spaced ring-shaped cracks encircling the glass fiber (Metcalfe et al. 1971; Hammami and Al-Ghuilani 2004). The formation of these hoop-like cracks could be attributed to the fluid-induced swelling of the polymeric phase or the swelling of the reaction products. It was also observed that flaws in glass fibers may occur in spiraling forms as well as in the longitudinal direction (Karbhari and Zhang 2003). Exposure of glass-reinforced polymeric composites to nitric acid resulted in crack initiation that was accompanied by subcritical growth, and subsequently followed by stable propagation. The first two stages were entirely random and confined to the fiber surfaces. The ensuing degradation of the composite was highly sensitive to the polymeric phase (Megel et al. 2001).

The aforementioned basic properties of glass fiber and the features of their interaction with fluid and stress have strong implications on the behavior of glass-reinforced composites. A comprehensive and detailed review that contains a wealth of technical details and an extensive bibliography on the subject was presented fairly recently (Schutte 1994). A discussion of the effects of fluids on damage, strength, and durability of glass fiber and glass fiber composites is contained in Sect. 7.1.

9.4 Galvanic Corrosive Effects

Carbon fibers have a different electrochemical potential than most metals, especially aluminum and steel. When placed in contact with each other, the carbon fiber/metal pairing gives rise to galvanic corrosion (Gebhard et al. 2009). This corrosion is induced by electrical currents that are generated by ion motions between the carbon fiber composites in contact with the metal. The phenomenon is amplified in the presence of moisture and sea water.

Galvanic corrosion may activate interfacial debonding between graphite fiber and polymers as well as induce internal fiber cracks (Gebhard et al. 2009). In different circumstances, it could corrode the metal.

Other forms of damage include changes in the chemical composition of the composite, rust formation and blistering within the composite, and pitting corrosion of the metal (Tucker et al. 1990).

Additional comprehensive lists of references are available elsewhere (Boyd et al. 1991; Boyd and Speak 1992; Fovet et al. 2000).

References

Adamson MJ (1980) Thermal expansion and swelling of cured epoxy resin used in graphite/epoxy composite materials. J Mater Sci 15(7):1736–1745

Apicella A, Nicolais L, Astarita G, Drioli E (1979) Effect of thermal history on water sorption, elastic properties and the glass transition of epoxy resions. Polymer 20:1143–1148

Boyd J, Speak S (1992) Galvanic corrosion effects on carbon fiber composites: results from accelerated tests. In 37th International SAMPE symposium and exhibition: materials working for you in the 21st century, 9–12 March 1992, pp 1184–1198

Boyd J, Chang G, Webb W, Speak S (1991) Galvanic corrosion effects on carbon fiber composites. In 36th International SAMPE symposium and exhibition: how concept becomes reality, 15–18 April 1991, pp 1217–1231

Charles RJ (1958a) Static fatigue of glass. I. J Appl Phys 29(11):1549–1553

Charles RJ (1958b) Static fatigue of glass. II. J Appl Phys 29(11):1554–1560

Fovet Y, Pourreyron L, Gal J (2000) Corrosion by galvanic coupling between carbon fiber posts and different alloys. Dent Mater 16(5):364–373

Gao J, Weitsman YJ (1998) Composites in the sea: sorption, strength and fatigue. University of Tennessee Report MAES 98-4.0 CM, August 1998

Gebhard A, Bayerl T, Schlarb AK, Friedrich K (2009) Galvanic corrosion of polyacrylnitrile (PAN) and pitch based short carbon fibres in polyetheretherketone (PEEK) composites. Corros Sci 51(11):2524–2528

Hammami A, Al-Ghuilani N (2004) Durability and environmental degradation of glass-vinylester composites. Polym Compos 25(6):609–616

Karbhari VM, Zhang S (2003) E-glass/vinylester composites in aqueous environments – I: experimental results. Appl Compos Mater 10(1):19–48

Megel M, Kumosa L, Ely T, Armentrout D, Kumosa M (2001) Initiation of stress-corrosion cracking in unidirectional glass/polymer composite materials. Compos Sci Technol 61 (2):231–246

Metcalfe AG, Schmitz GK (1972) Mechanism of stress corrosion in E-glass filaments. Glass Technol 13(1):5–16

Metcalfe AG, Gulden ME, Schmitz GK (1971) Spontaneous cracking of glass filaments. Glass Technol 12(1):15–23

Prian L, Barkatt A (1999) Degradation mechanism of fiber-reinforced plastics and its implications to prediction of long-term behavior. J Mater Sci 34(16):3977–3989

Regester RF (1969) Behavior of fiber reinforced plastic materials in chemical service. Corrosion 25(4):157–167

Schutte CL (1994) Environmental durability of glass-fiber composites. Mater Sci Eng R Rep 13 (7):265–323

Sheard PA, Jones FR (1986) The stress corrosion of glass fibres andepoxy composites in aqueous environments. In Loo TT, Sun CT (eds) Proceedings of the international symposium on composite materials and structures, Beijing, 10–13 June 1986, pp 118–123

Suh D, Ku M, Nam J, Kim B, Yoon S (2001) Equilibrium water uptake of epoxy/carbon fiber composites in hygrothermal environmental conditions. J Compos Mater 35(3):264–278

Tucker WC, Brown R, Russell L (1990) Corrosion between a graphite/polymer composite and metals. J Compos Mater 24(1):92–102

Vieth WR, Sladek KJ (1965) A model for diffusion in a glassy polymer. J Colloid Sci 20 (9):1014–1033

Vieth W, Howell J, Hsieh J (1976) Dual sorption theory. J Memb Sci 1:177–220

Weitsman YJ (1990) A continuum diffusion model for viscoelastic materials. J Phys Chem 94 (2):961–968

Chapter 10
Concluding Remarks

Fluids are a stealthy substance and even though in many circumstances their effects may be of secondary importance, the disregarding of their ingress and presence can lead to perilous consequences.

The very nature of the phenomenon precludes the establishment of comprehensive rules and requires a case by case study. Nevertheless, it is possible to propose some basic guidelines for assessing the effects of fluids on polymers and polymeric composites and list the most prominent circumstances when those effects should be of concern. These are noted below.

- A preliminary assessment of the applicability of a certain material to operate in an intended environment can be achieved by exposure to that ambience and recording weight gains. Data along curves "C" or "D" sketched in Fig. 4.2 should raise doubts about the validity of the material choice.

  Note that while weight gain data are accelerated by temperature, one should never collect them at 20–30°C above that of the intended application to minimize synergistic effects.
- If long term exposure is contemplated, then coupling with the time dependent response of the polymer may be considered, employing concepts detailed in Chap. 6.
- It should be realized that more damage is caused by exposure to cyclic environment than by exposure to constant ambience.
- Joints and free edges are the most vulnerable sites for stealthy ingress of ambient fluids as may be inferred from Fig. 10.1 (Earl et al. 2003). Such locations require special and detailed attention. The joining of composites to metal may trigger a corrosive galvanic effect.
- Dimensional stability may be strongly affected by the ingress of fluids, especially in thin walled composite structures and in adhesive joints. Typically, the effect of 1% fluid weight gain is tantamount to a thermal excursion of 100°C.

  An illustration of the effects of daily fluctuations in ambient moisture was provided recently (Siriruk 2009). That example considered $[90°/0°]_S$ layup of gr/ep exposed to monthly fluctuations of ambient moisture $0 < m < 1\%$ at its

Y.J. Weitsman, *Fluid Effects in Polymers and Polymeric Composites*, Mechanical Engineering Series, DOI 10.1007/978-1-4614-1059-1_10,

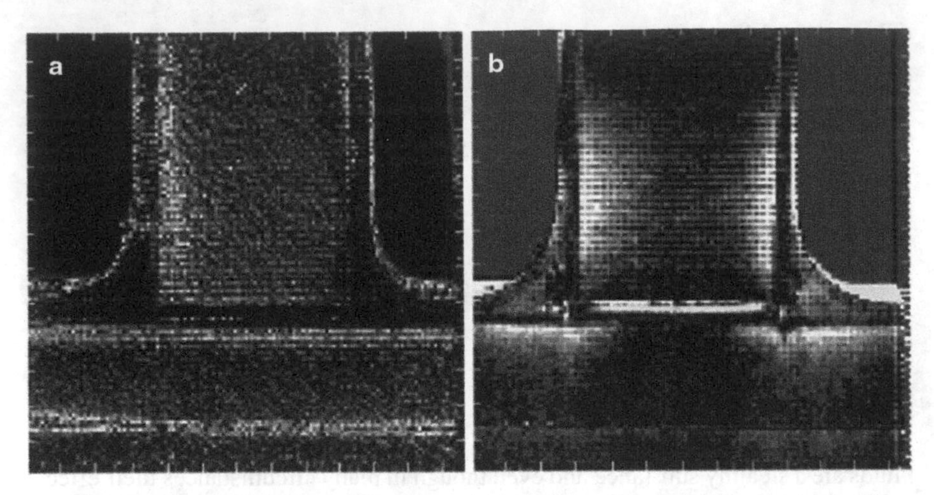

Fig. 10.1 Full-field DeltaTherm readings from (**a**) a wet joint surface and (**b**) a dry joint surface. Reprinted from Earl et al. (2003), Copyright (2003), with permission from Elsevier

outer boundary and maintained at $m = 0$ at the interior boundary. It was shown that the corresponding curvature k of the laminate varied between 0 and 0.25 (1/m). Such curvature may sufficiently distort the shape of the laminate in case that high precision is important.

- Fatigue life may be substantially reduced in the presence of fluids, especially under immersed circumstances. Failure modes differ materially from those of the dry case and may lead to abrupt failures. Such reductions are also likely to occur in glass fiber reinforced composites due to enhanced fiber failures.

The above effects can be instigated by the presence of undetectable damage.

References

Earl JS, Dulieu-Barton JM, Shenoi RA (2003) Determination of hygrothermal ageing effects in sandwich construction joints using thermoelastic stress analysis. Compos Sci Technol 63 (2):211–223

Siriruk A (2009) The mechanical characterization of polymeric sandwich materials for marine applications. Dissertation (Ph. D.), University of Tennessee, Knoxville

Index

Y.J. Weitsman, *Fluid Effects in Polymers and Polymeric Composites*, Mechanical Engineering Series, DOI 10.1007/978-1-4614-1059-1,